Stratigraphic Traps in Carbonate Rocks

Selected papers reprinted from
AAPG Bulletin and Special Publications

compiled by
Dr. S. J. Mazzullo

Published May 1980 by
The American Association of Petroleum Geologists
Tulsa, Oklahoma, U.S.A.

Printed by
Edwards Brothers, Inc.
Ann Arbor, Michigan, U.S.A.

Contents:

Preface:

Carbonate Facies:

It is a relatively simple assertion, yet one that is profound in its application: the subsurface location of subtle, stratigraphic-type hydrocarbon accumulations must rely on detailed facies analyses in conjunction with conventional geological-geophysical methods. Indeed, most petroleum geologists are aware not only of the potential for (as yet) undiscovered hydrocarbon reserves in stratigraphic traps, but also of the stark reality of ever-decreasing numbers of available structural plays in mature exploration provinces.

The significance to petroleum geologists of stratigraphic traps is further documented by the various AAPG Memoirs and other publications, including those edited by Levorsen (1941), Weeks (1958), Halbouty (1970), King (1972), Braunstein (1976), and Payton (1977). Stratigraphic traps in siliciclastic rocks have been given their due in AAPG Reprint Series Numbers 7, 8 (Weimer, 1973), and 20 (Ali and Friedman, 1977). This Reprint Series makes the effort for equal time by presenting 10 previously published papers that deal with stratigraphically-trapped hydrocarbon accumulations in carbonate reservoirs.

The inclusion of these specific papers, from among the 90 or so published in past years by AAPG (see Appendix B), was a choice dictated by my objectives: (1) to illustrate a variety of carbonate stratigraphic-type traps and associated facies (depositional and/or diagenetic) at the field-study level, and; (2) to provide a reference for those petroleum geologists who may wish to personally perform field-stratigraphic analyses. Accordingly, I have purposely omitted references to regional studies or those field studies for which little lithostratigraphic information was available. The included papers describe and illustrate *rocks*, the essence of all geological investigations. The obvious skewness in geographic scope and reservoir ages of included papers is a function of my objectives-related selection process and the limiting guidelines of the Reprint Series format.

In order to illustrate stratigraphic traps, their geologic character must first be defined. From among the numerous schemes of reservoir classification (e.g., Levorsen, 1936, 1941, 1954; Martin, 1966; Sanders, 1943), that of Rittenhouse (1972) was considered most functional. Accordingly, many of the papers herein describe fields in which reservoir *occurrence* relates directly to depositional and/or depositionally-controlled diagenetic facies variations (facies-change traps); although there may be disagreement as to the precise classification of those specific fields in which regional or local structure may have been essential elements of hydrocarbon entrapment. Following these concepts, also included are outstanding examples of unconformity-related (paleogeomorphic) stratigraphic fields. Appendix A is a limited selection of stratigraphic-type carbonate fields from numerous basins, replete with information on the range of ages and paleodepositional environments represented by the individual reservoirs. Also included are some references to fields in which the stratigraphic controls on hydrocarbon accumulation are accentuated by significant structuring (e.g., Poza Rica trend).

The papers appear in reservoir-chronological order, and each illustrates the details of facies control on reservoir occurrence, with examples of shallow-water "reefs" and grainstone shoals, dolomitized inner-shelf facies, and basinal deposits. The prolific Devonian reef trend fields of Canada are exhumed for analysis by Hemphill, Smith, and Szabo (1970), whereas Gray (1967) and Toomey and Winland (1973) offer in-depth profiles of shelf-edge, phylloid algal complexes from Pennsylvanian reservoirs of the Paradox and Permian basins, respectively. Thomas and Glaister (1960) and Hansen (1966) illustrate the Mississippian reef and grainstone reservoirs of the Williston basin, some of which are associated with paleogeomorphic traps. Terriere (1976) and Tyler and Erwin (1976) provide examples of shelf-edge, bioherm-complex reservoirs from the Cretaceous trends of the Gulf Coast (US) province; Coogan, Bebout, and Maggio (1972) present their views of the depositional environments represented by the immense Cretaceous Golden Lane and Poza Rica trends of Mexico. Ottmann, Keyes, and Zieglar (1976) describe dolomitized shoreline carbonate reservoirs and associated facies from the Jurassic Jay field area (Smackover) of Florida. Finally, Bebout and Pendexter (1975) highlight the significance of porosity evolution to reservoir formation in their discussions of Tertiary shelf limestones of Zelten field in Libya.

I would like to acknowledge Gary Howell and Ron Hart of AAPG for their continuing interest in this Reprint. I also thank Herb Rehders, Miles Gray, Angela Sanchez, and Fawn Meek (Union Texas Petroleum, Midland), for their invaluable assistance in gathering production data (Appendix A) and typing.

S. J. Mazzullo
Union Texas Petroleum
Midland, Texas
May, 1980

References Cited

Ali, S., and G. M. Friedman, eds., 1977, Diagenesis of sandstones: AAPG Reprint Series 20, 239 p.

Braunstein, J., ed., 1976, North American oil and gas fields: AAPG Mem. 24, 360 p.

Halbouty, M. T., ed., 1970, Geology of giant petroleum fields: AAPG Mem. 14, 575 p.

King, R. E., ed., 1972, Stratigraphic oil and gas fields: AAPG Mem. 16, 687 p.

Levorsen, A. I., 1936, Stratigraphic versus structural accumulation: AAPG Bull., v. 20, p. 521-530.

—— ed., 1941, Stratigraphic type oil fields: Tulsa, AAPG, 902 p.

—— 1954, Geology of petroleum: San Francisco, W. H. Freeman & Co., 703 p.

Martin, R., 1966, Paleogeomorphology and its application to exploration for oil and gas (with examples from Western Canada): AAPG Bull., v. 50, p. 2277-2311.

Payton, C. E., ed., 1977, Seismic stratigraphy—applications to hydrocarbon exploration: AAPG Mem. 26, 516 p.

Rittenhouse, G., 1972, Stratigraphic-trap classification, in R. E. King, ed., Stratigraphic oil and gas fields: AAPG Mem. 16, p. 14-28.

Sanders, C. W., 1943, Stratigraphic type oil fields and proposed traps: AAPG Bull., v. 27, p. 539-550.

Weeks, L. G., ed., 1958, Habitat of oil—a symposium: Tulsa, AAPG, 1384 p.

Weimer, R. J., ed., 1973, Sandstone reservoirs and stratigraphic concepts I, II: AAPG Reprint Series 7, 8, 190 p., 216 p.

Appendix A
Selected Carbonate Stratigraphic Traps[1]

Field Name	Location[2]	Age	Dimensions (miles)	Reservoir — Depositional Environment	Reserves[3] Oil (MMBO)	Reserves[3] Gas (BCF)	Discovery Methods
Adair	PB	Permian	2.5x5	Atoll bank complex	21.5	--	Seismic
Alida-Nottingham Trend	WB	Miss.	1.5x4*	Oolite bars (subunconformity)	500	--	Seismic
Aneth	PxB	Penn.	12x12	Shelf-edge algal bioherm complex	375	--	Seismic
Azalea	PB	Penn.	1x13	Shelf-edge algal bank complex	13**	--	Seismic
Black Lake	SU	Early Cret.	5x8	Shelf-edge bio-oolite bank	156	892	Seismic, subsurface
Block 42	PB	Penn.	2x6	Lagoonal algal bioherm complex	7.8**	--	Seismic
Bonnie Glen	AB	Late Dev.	5x5	Shelf/edge bioherm	335	--	Seismic
Boyd-Peters	MB	Silurian	1x1	Pinnacle bioherms	5.9**	0.29**	Gravity
Carthage	SU	Early Cret.	24x48	Shelf-edge oolite bars	50**	4,000**	Seismic
Chapman Deep	PB	Penn.	1x2.5	Shelf-edge algal-oolite bank	--	15	Seismic
Chaveroo	PB	Permian	6x9	Peritidal dolomite	20**	--	Seismic
Cottonwood Creek	BHB	Permian	4x8	Oolite bar-beach	182	--	Subsurface
Credo	PB	Permian	3x3	Shelf-edge bioclastic bank	11	3.9	Seismic
Eagle Springs	BR	Te,To	1.5x2	Lacustrine	10	--	Seismic
Empire Abo	PB	Permian	2x12	Dolomite shelf-edge bioherm(?)	250	--	Subsurface
Fairway	ET	Early Cret.	6x9	Shelf/edge bioherm complex	200	475	Seismic
Ft. Terrett Ranch	PB	Penn.	3x3	Inner-shelf bioherm(?)	10	--	--
Glenburn	WB	Miss.	2x4	Algal bioherm, oolite	13.3	--	Seismic
Glenevis	AB	Miss.	1x3	Dolomite (subunconformity-peritidal)	22	--	Seismic
Golden Lane Trend	TE	Middle Cret.	1.3*x110	Shelf-edge rudistid bioherm	1,100**	--	Oil seeps
Goose Lake	WB	Miss.	1.5x8	Shelf algal bioherm	9.6**	--	Seismic
Greenwood	AkB	Penn.	--	Bioclastic-oolite shoals	--	1,000	Seismic, subsurface
Haldimand	AA	Middle Sil.	--	Peritidal (?) dolomite	--	126	--
Hallanan	PB	Penn.	1.5x6.5	Bioclastic beach-bar	4.7**	3.9**	Chance
Harper	PB	Permian	2x2	Peritidal dolomite	33**	12.3**	Subsurface
Hugoton	AkB	Permian	45x150	Peritidal dolomite	--	60,000	Surface
Hulldale	PB	Penn.	3x5	Shelf-edge bioherm complex	37	30.8	Seismic
Idris "A"	SB	Tp	4x8	Shelf bioherm	2,000	--	Seismic
Jameson Reef Trend	PB	Penn.	6x12	Shelf-edge algal bioherm complex	45**	43.1**	Seismic
Jay	WF/PA	Jurassic	4x7	Peritidal dolomite pelgrainstone	346	350	Seismic
Joffre-1	AB	Late Dev.	3x5	Shelf limestone (?facies)	70	--	Seismic
Kaybob	PRA	Late Dev.	3x10	Shelf-edge bioherm complex	300	--	Seismic
Kingdom Abo	PB	Permian	1.5x20	Shelf-edge dolomite bioherm (?)	15	--	Seismic, subsurface
Kemnitz	PB	Permian	2.5x4.5	Shelf-edge algal bioherm	14**	--	Seismic
Leduc-Woodbend	AB	Late Dev.	10x16	Shelf-edge bioherm	500	--	Seismic

Appendix A. Continued

Field	Basin	Age	Dimensions	Reservoir Type	Reserve	Reserve	Detection
Lincoln SE	NU	Penn.	4x5	Shelf-edge oolite bars	18	—	Subsurface
Marine	IB	Sil.-Dev.	5x5	Suprareef biocalcarenite (shelf)	12(+)	—	Seismic
Michigan Reef Trend	MB	Silurian	1.5x1.5*	Pinnacle bioherms	600	—	Seismic, subsurface
Millican	PB	Penn.	1.5x4	Shelf-edge bioherm complex	8**	12.0**	Seismic
Nena Lucia	PB	Penn.	1.3x11	Shelf-edge algal bioherm	40	—	—
Newburg-So. Westhope	WB	Miss.	4x13	Bio-oolite grainstone (subunconformity)	27	—	Subsurface
No. Knox City	PB	Penn.	1x5	Platform algal bioherm	10.5**	—	Seismic
Page	PB	Penn.	2x4	Shelf-edge bioherm complex	6.0**	1.0**	Seismic
Poza Rica	TE	Middle Cret.	6x17	Gravity-debris flows (Tamabra)	2,100	—	Surface, seismic
Putnam	AkB	Penn.	2-4x36	Shelf-edge bioherm	16	534	Chance
Rainbow Area	BCB	Middle Dev.	2x2*	Shelf/edge pinnacle-atoll	720	—	Seismic
Redwater	PRA	Late Dev.	20x20	Shelf-edge bioherms	817	—	Seismic
Rival-Lignite	WB	Miss.	5x8	Shelf algal bank	25	—	Seismic
Scurry	PB	Ord.-Perm.	5x24	Atoll bank complex	172,000	—	Seismic
Seminole SE	PB	Penn.	1x2	Shelf-edge coral-grainstone bank	1.7	3.0	Seismic
Slaughter-Levelland	PB	Permian	20x30	Progradational oolite shoals	700**	360**	Subsurface, seismic
Star-Lacey	AkB	Sil.-Dev.	2x8	Shelf dolomite	11.7	30(+)	Seismic
Sunoco-Felda	SFE	Early Cret.	3x4	Shelf/edge algal-gastropod mound	44	—	Subsurface, seismic
Swan Hills	PRA	Late Dev.	20x33	Shelf/edge bioherm complex	999	—	Seismic
Tocito Dome	PxB	Penn.	2x5.5	Shelf algal-coral bioherm	12(+)	—	Structure
Velma	PB	Permian	2x2	Lagoonal algal bioherm complex	2	2.5	Chance
Walker Creek	NL	Late Jurassic	1-3x11	Inner-shelf oolite bank	96	100	Seismic
Wellman	PB	Permian	2x2	Atoll algal bioherm complex	26**	6.0**	Seismic
West Campbell	AkB	Sil.-Dev.	5.5x6.5	Peritidal dolomite (subunconformity)	0.25	84,000	Seismic
Zama	BCB	Middle Dev.	10x18	Shelf-edge bioherm	Classified	Classified	Seismic
Zelten	SB	Tp-Te	8x17	Bioclastic-coralgal banks	1,500**	—	—

[1]The data presented herein were obtained from the references cited in Appendix B, as well as from numerous regional publications and industry sources. Reserve estimates for some of the larger oil and gas fields/trends are from Halbouty in AAPG Memoir 14 (1970).

[2](Location Guide): (AA) Algonquin Arch.; (AB) Alberta Basin; (AkB) Anadarko Basin; (BCB) Black Creek Basin; (BHB) Big Horn Basin; (BR) Basin and Range (Nevada); (CA) Cambridge Arch.; (ET) East Texas; (GC) Gulf Coast; (IB) Illinois Basin; (MB) Michigan Basin; (NL) No. Louisiana Salt Basin; (NU) Nemaha Uplift; (PB) Permian Basin; (PRA) Peace River Arch.; (PxB) Paradox Basin; (SB) Sirte Basin (Libya); (SFE) South Florida Embayment; (SU) Sabine Uplift; (TE) Tampico Embayment; (WB) Williston Basin; (WF/PA) West Florida/Pensacola Arch.

[3]Reserve estimates are (believed to be) ultimate recoverable, as per author source; numbers denoted by a double asterisk represent most recent available production-to-date figures (from author or industry sources). For many specific fields, only oil or gas data were provided in the source.

* Denotes average dimensions of individual fields within the trend.

** Denotes most recent production-to-date figures (from author or industry sources).

Appendix B:
Selected Bibliography

AAPG Sources

Akin, R. H., and R. W. Graves, 1969, Reynolds Oolite of southern Arkansas: AAPG Bull., v. 53, p. 1909-1922.

Andrichuk, J. M., 1958, Cooking Lake and Duvernay (Late Devonian) sedimentation in Edmonton area of Central Alberta, Canada: AAPG Bull., v. 42, p. 2189-2222.

—— 1958, Mississippian Madison stratigraphy and sedimentation in Wyoming and southern Montana, in L. G. Weeks, ed., Habitat of oil: a symposium: Tulsa, AAPG, p. 225-267.

—— 1958, Stratigraphy and facies analysis of Upper Devonian reefs in Leduc, Stettler, and Redwater areas, Alberta: AAPG Bull., v. 42, p. 1-93.

—— 1960, Facies analysis of Upper Devonian Wabamun Group in west-central Alberta, Canada: AAPG Bull., v. 44, p. 1651-1681.

—— 1961, Stratigraphic evidence for tectonic and current control of Upper Devonian reef sedimentation, Duhamel area, Alberta, Canada: AAPG Bull., v. 45, p. 612-632.

Ball, M. M., 1972, Exploration methods for stratigraphic traps in carbonate rocks, in R. E. King, ed., Stratigraphic oil and gas fields: AAPG Mem. 16, p. 64-81.

Barss, D. L., A. B. Copland, and W. D. Ritchie, 1970, Geology of Middle Devonian reefs, Rainbow areas, Alberta, Canada, in M. T. Halbouty, ed., Geology of giant petroleum fields: AAPG Mem. 14, p. 19-49.

Beales, F. W., and A. E. Oldershaw, 1969, Evaporite-solution brecciation and Devonian carbonate reservoir porosity in Western Canada: AAPG Bull., v. 53, p. 503-512.

Bebout, D. G., and C. Pendexter, 1975, Secondary carbonate porosity as related to Early Tertiary depositional facies, Zelten field, Libya: AAPG Bull., v. 59, p. 665-693.

Bergenback, R. E., and R. T. Terriere, 1953, Petrography and petrology of Scurry Reef, Scurry County, Texas: AAPG Bull., v. 37, p. 1014-1029.

Bishop, W. F., 1968, Petrology of Upper Smackover Limestone in North Haynesville field, Claiborne Parish, Louisiana: AAPG Bull., v. 52, p. 92-128.

—— 1971, Geology of a Smackover stratigraphic trap: AAPG Bull., v. 55, p. 51-63.

—— 1971, Geology of upper member of Buckner Formation, Haynesville field area, Claiborne Parish, Louisiana: AAPG Bull., v. 55, p. 566-580.

Bubb, J. N., and W. G. Hatlelid, 1977, Seismic stratigraphy and global changes of sea level, Part 10: seismic recognition of carbonate build-ups, in C. E. Payton, ed., Seismic stratigraphy—applications to hydrocarbon exploration: AAPG Mem. 26, p. 185-204.

Chimene, C. A., 1976, Upper Smackover reservoirs, Walker Creek field area, Lafayette and Columbia Counties, Arkansas, in J. Braunstein, ed., North American oil and gas fields: AAPG Mem. 24, p. 177-204.

Coogan, A. H., D. G. Bebout, and C. Maggio, 1972, Depositional environments and geologic history of Golden Lane and Poza Rica trend, Mexico, an alternative view: AAPG Bull., v. 56, p. 1419-1447.

Cussey, R., and G. M. Friedman, 1977, Patterns of porosity and cement in ooid reservoirs in Dogger (Middle Jurassic) of France: AAPG Bull., v. 61, p. 511-518.

Doss, A. K., and F. J. Spiva, 1968, Palo Pinto Limestone production, western Runnels County, Texas: AAPG Bull., v. 52, p. 82-91.

Downing, J. A., and D. Y. Cooke, 1955, Distribution of reefs of Woodbend Group in Alberta, Canada: AAPG Bull., v. 39, p. 189-206.

Ebanks, W. J., R. M. Euwer, and D. E. Nodine-Zeller, 1977, Mississippian combination trap, Bindley field, Hodgeman County, Kansas: AAPG Bull., v. 61, p. 309-330.

Edie, R. W., 1958, Mississippian sedimentation and oil fields in southeastern Saskatchewan: AAPG Bull., v. 42, p. 94-126 (also in A. J. Goodman, ed., 1958, Jurassic and Carboniferous of Western Canada: Tulsa, AAPG, p. 331-363).

—— 1959, Middle Devonian sedimentation and oil possibilities, Central Saskatchewan, Canada: AAPG Bull., v. 43, p. 1026-1057.

El-Naggar, Z. R., 1973, Stratigraphy and microfacies of type Magwa Formation of Kuwait, Arabia, pt. 2: Mishrif Limestone Member: AAPG Bull., v. 57, p. 2263-2279.

Evans, H., 1972, Zama—a geophysical case history, in R. E. King, ed., Stratigraphic oil and gas fields: AAPG Mem. 16, p. 440-452.

Ferris, C., 1972, Boyd-Peters Reef, St. Clair County, Michigan, in R. E. King, ed., Stratigraphic oil and gas fields: AAPG Mem. 16, p. 460-471.

Fong, G., 1960, Geology of Devonian Beaverhill Lake Formation, Swan Hills area, Alberta, Canada: AAPG Bull., v. 44, p. 195-209.

Freeman, L. B., 1941, Big Sinking field, Lee County, Kentucky, in A. I. Levorsen, ed.,

Stratigraphic type oil fields: Tulsa, AAPG, p. 166-207.

Fuller, J. G. C. M., and J. W. Porter, 1969, Evaporite formations with petroleum reservoirs in Devonian and Mississippian of Alberta, Saskatchewan, and North Dakota: AAPG Bull., v. 53, p. 909-926.

Garlough, J. L., and G. L. Taylor, 1941, Hugoton gas field, Grant, Haskell, Morton, Stevens and Seward Counties, Kansas, and Texas County, Oklahoma, in A. I. Levorsen, ed., Stratigraphic type oil fields: Tulsa, AAPG, p. 78-104.

Gill, D., 1979, Differential entrapment of oil and gas in Niagaran pinnacle-reef belt of northern Michigan: AAPG Bull., v. 63, p. 608-620.

Goodman, A. J., 1945, Limestone reservoir conditions in Turner Valley oil field, Alberta, Canada: AAPG Bull., v. 29, p. 1156-1168.

Gray, R. S., 1967, Cache field—a Pennsylvanian algal reservoir in southwestern Colorado: AAPG Bull., v. 51, p. 1959-1978.

Hansen, A. R., 1966, Reef trends of Mississippian Ratcliffe Zone, northeast Montana and northwest North Dakota: AAPG Bull., v. 50, p. 2260-2268.

Harris, S. A., 1975, Hydrocarbon accumulation in "Meramac-Osage" (Mississippian) rocks, Sooner Trend, northwest-central Oklahoma: AAPG Bull., v. 59, p. 633-664.

Harris, S. H., C. B. Land, and J. H. McKeever, 1966, Relation of Mission Canyon stratigraphy to oil production in north-central North Dakota: AAPG Bull., v. 50, p. 2269-2276.

Harvey, R. L., 1972, West Campbell (Northwest Cedardale) gas field, Major County, Oklahoma, in R. E. King, ed., Stratigraphic oil and gas fields: AAPG Mem. 16, p. 568-578.

Harwell, J. C., and W. R. Rector, 1972, North Knox City field, Knox County, Texas, in R. E. King, ed., Stratigraphic oil and gas fields: AAPG Mem. 16, p. 453-459.

Hemphill, C. R., R. I. Smith, and F. Szabo, 1970, Geology of Beaverhill Lake reefs, Swan Hills area, Alberta, in M. T. Halbouty, ed., Geology of giant petroleum fields: AAPG Mem. 14, p. 50-90.

Hemsell, C. C., 1939, Geology of Hugoton gas field of southwestern Kansas: AAPG Bull., v. 23, p. 1054-1067.

Herman, G., and C. A. Barkell, 1957, Pennsylvanian stratigraphy and productive zones, Paradox Salt basin: AAPG Bull., v. 41, p. 861-881.

Hriskevich, M. E., 1970, Middle Devonian reef production, Rainbow area, Alberta, Canada: AAPG Bull., v. 54, p. 2260-2281.

Hull, C. E., and H. R. Warman, 1970, Asmari oil fields of Iran, in M. T. Halbouty, ed., Geology of giant petroleum fields: AAPG Mem. 14, p. 428-437.

Imholz, H. W., 1941, Noodle Creek pool, Jones County, Texas, in A. I. Levorsen, ed., Stratigraphic type oil fields: Tulsa, AAPG, p. 698-721.

Jenik, A. J., and J. F. Lerbekmo, 1968, Facies and geometry of Swan Hills reef member of Beaverhill Lake Formation (Upper Devonian), Goose River field, Alberta, Canada: AAPG Bull., v. 52, p. 21-56.

Jodry, R. L., 1969, Growth and dolomitization of Silurian reefs, St. Clair County, Michigan: AAPG Bull., v. 53, p. 957-981.

Klovan, J. E., 1974, Development of Western Canadian Devonian reefs and comparison with Holocene analogues: AAPG Bull., v. 58, p. 787-799.

Langton, J. R., and G. E. Chin, 1968, Rainbow Member facies and related reservoir properties, Rainbow Lake, Alberta: AAPG Bull., v. 52, p. 1925-1955.

Layer, D. B., et al, 1949, Leduc oil field, Alberta, a Devonian coral-reef discovery: AAPG Bull., v. 33, p. 572-602.

LeMay, W. J., 1972, Empire Abo field, southeast New Mexico, in R. E. King, ed., Stratigraphic oil and gas fields: AAPG Mem. 16, p. 472-480.

Lowenstam, H. A., 1948, Marine pool, Madison County, Illinois, Silurian reef producer, in J. V. Howell, ed., Structure of typical American oil fields, v. III: Tulsa, AAPG, p. 153-188.

Malek-Aslani, M., 1970, Lower Wolfcampian reef in Kemnitz field, Lea County, New Mexico: AAPG Bull., v. 54, p. 2317-2335.

Marafi, H., 1972, Newburg-South Westhope oil fields, North Dakota, in R. E. King, ed., Stratigraphic oil and gas fields: AAPG Mem. 16, p. 633-640.

McCamis, J. G., and L. S. Griffith, 1968, Middle Devonian facies relations, Zama area, Alberta: AAPG Bull., v. 52, p. 1899-1924.

Mesolella, K. J., J. D. Robinson, L. M. McCormick, and A. R. Ormiston, 1974, Cyclic deposition of Silurian carbonates and evaporites in Michigan Basin: AAPG Bull., v. 58, p. 34-62.

Metwalli, M. H., and Y. E. Abd El-Hady, 1975, Petrographic characteristics of oil-bearing rocks in Alamein oil field; significance in source-reservoir relations in northern Western Desert, Egypt: AAPG Bull., v. 59, p. 510-523.

Miller, E. G., 1972, Parkman field, Williston basin, Saskatchewan, in R. E. King, ed., Stratigraphic oil and gas fields: AAPG Mem. 16, p. 502-510.

Moore, G. T., 1973, Lodgepole limestone facies in

southwestern Montana: AAPG Bull., v. 57, p. 1703-1713.

Murray, D. K., and L. C. Bortz, 1967, Eagle Springs oil field, Railway Valley, Nye County, Nevada: AAPG Bull., v. 51, p. 2133-2145.

Murphy, J. K., P. E. M. Purcell, and H. E. Barton, 1941, Seymour pool, Baylor County, Texas, in A. I. Levorsen, ed., Stratigraphic type oil fields: Tulsa, AAPG, p. 760-775.

Ottmann, R. D., P. L. Keyes, and M. A. Zieglar, 1976, Jay field, Florida—a Jurassic stratigraphic trap, in J. Braunstein, ed., North American oil and gas fields: AAPG Mem. 24, p. 276-286.

Pedry, J. J., 1957, Cottonwood Creek field, Washakie County, Wyoming, carbonate stratigraphic trap: AAPG Bull., v. 41, p. 823-838.

Peterson, J. A., 1966, Stratigraphic vs. structural controls on carbonate-mound hydrocarbon accumulation, Aneth area, Paradox basin: AAPG Bull., v. 50, p. 2068-2081.

——— and R. J. Hite, 1969, Pennsylvanian evaporite-carbonate cycles and their relation to petroleum occurrence, southern Rocky Mountains: AAPG Bull., v. 53, p. 884-908.

Picard, M. D., 1959, White Mesa field, environmental trap, San Juan County, Utah: AAPG Bull., v. 43, p. 2456-2469.

——— B. R. Brown, A. J. Loleit, and J. W. Parker, 1960, Geology of Pennsylvanian gas in Four Corners region: AAPG Bull., v. 44, p. 1541-1569.

Pippin. L., 1970, Panhandle-Hugoton field, Texas-Oklahoma-Kansas—The first fifty years, in M. T. Halbouty, ed., Geology of giant petroleum fields: AAPG Mem. 14, p. 204-222.

Procter, R. M., and G. MacAuley, 1968, Mississippian of western Canada and Williston basin: AAPG Bull., v. 52, p. 1956-1968.

Rockwell, D. W., and A. G. Rojas, 1953, Coordination of seismic and geologic data in Poza Rica-Golden Lane area, Mexico: AAPG Bull., v. 37, p. 2551-2565.

Roehl, P. O., 1967, Stony Mountain (Ordovician) and Interlake (Silurian) facies analogs of recent low-energy marine and subaerial carbonates: AAPG Bull., v. 51, p. 1979-2032.

Rogers, R. E., 1968, Carthage field, Panola County, Texas, in B. W. Beebe, ed., Natural gases of North America: AAPG Mem. 9, p. 1020-1059.

Salas, G. P., 1949, Geology and development of Poza Rica oil field, Veracruz, Mexico: AAPG Bull., v. 33, p. 1385-1409.

Scholle, P. A., 1977, Chalk diagenesis and its relation to petroleum exploration; Oil from chalks, a modern miracle?: AAPG Bull., v. 61, p. 982-1009.

Schultheis, N. H., 1976, Kaybob oil field, Alberta, Canada, in J. Braunstein, ed., North American oil and gas fields: AAPG Mem. 24, p. 79-90.

Schwalb, and E. N. Wilson, 1972, Greenburg Consolidated oil pool, Green and Taylor Counties, Kentucky, in R. E. King, ed., Stratigraphic oil and gas fields: AAPG Mem. 16, p. 579-584.

Sharma, G. D., 1966, Geology of Peters Reef, St. Clair County, Michigan: AAPG Bull., v. 50, p. 327-350.

Silver, B. A., and R. G. Todd, 1969, Permian cyclic strata, northern Midland and Delaware basins, West Texas and southeastern New Mexico: AAPG Bull., v. 53, p. 2223-2251.

Smith, F. W., and G. E. Summers, D. Wallington, and J. L. Lee, 1958, Mississippian oil reservoirs in Williston basin, in L. G. Weeks, ed., Habitat of oil: a symposium: Tulsa, AAPG, p. 149-177.

Stout, J. L., 1964, Pore geometry as related to carbonate stratigraphic traps: AAPG Bull., v. 48, p. 329-337.

Terriere, T. R., 1976, Geology of Fairway field, East Texas, in J. Braunstein, ed., North American oil and gas fields: AAPG Mem. 24, p. 157-176.

Thomas, G. E., and R. P. Glaister, 1960, Facies and porosity relationships in some Mississippian carbonate cycles of Western Canada basin: AAPG Bull., v. 44, p. 569-588.

Thornton, D. E., and H. H. Gaston, 1968, Geology and development of Lusk Strawn field, Eddy and Lea Counties, New Mexico: AAPG Bull., v. 52, p. 66-81.

Todd, R. G., 1976, Oolite-bar progradation, San Andres Formation, Midland basin, Texas: AAPG Bull., v. 60, p. 907-925.

Toomey, D. F., and H. D. Winland, 1973, Rock and biotic facies associated with Middle Pennsylvanian (Desmoinesian) algal buildup, Nena Lucia field, Nolan County, Texas: AAPG Bull., v. 57, p. 1053-1074.

Towse, D., 1957, Petrology of Beaver Lodge Madison Limestone reservoir, North Dakota: AAPG Bull., v. 41, p. 2493-2507.

Tyler, A. N., and W. L. Erwin, 1976, Sunoco-Felda field, Hendry and Collier Counties, Florida, in J. Braunstein, ed., North American oil and gas fields: AAPG Mem. 24, p. 287-299.

Vest, E. L., 1970, Oil fields of Pennsylvanian-Permian Horseshoe Atoll, West Texas, in M. T. Halbouty, ed., Geology of giant petroleum fields: AAPG Mem. 14, p. 185-203.

Viniegra, F. O., and C. Castillo-Tejero, 1970, Golden Lane fields, Veracruz, Mexico, in M. T. Halbouty, ed., Geology of giant petroleum fields: AAPG Mem. 14, p. 309-325.

Wardlaw, N. C., and G. E. Reinson, 1971, Carbonate and evaporite deposition and diagenesis, Middle Devonian Winnipegosis and Prairie Evaporite Formations of south-central Saskatchewan: AAPG Bull., v. 55, p. 1759-1786.

Waring, W. W., and D. B. Layer, 1950, Devonian dolomitized reef, D3 reservoir, Leduc field, Alberta, Canada: AAPG Bull., v. 34, p. 295-312.

White, B. R., 1972, Black Lake field, Natchitoches Parish, Louisiana, in R. E. King, ed., Stratigraphic oil and gas fields: AAPG Mem. 16, p. 481-488.

Wilson, H. H., 1977, "Frozen-in" hydrocarbon accumulations or diagenetic traps: Exploration targets: AAPG Bull., v. 61, p. 483-491.

Wingerter, H. R., 1968, Greenwood field, Kansas, Colorado, and Oklahoma, in B. W. Beebe, ed., Natural gases of North America: AAPG Mem. 9, p. 1557-1566.

Withrow, P. C., 1972, Star-Lacey field, Blaine and Kingfisher Counties, Oklahoma, in R. E. King, ed., Stratigraphic oil and gas fields: AAPG Mem. 16, p. 520-531.

Non-AAPG Sources

Badon, C. L., 1974, Petrology and reservoir potential of the upper member of the Smackover Formation, Clarke County, Mississippi: Gulf Coast Assoc. Geol. Socs. Trans., v. 24, p. 163-174.

Barnetche, A., and L. V. Illing, 1956, The Tamabra Limestone of the Poza Rica oil field, Veracruz, Mexico: 20th Int. Geol. Congress (Mexico), 38 p.

Bishop, W. F., 1971, Stratigraphic control of production from Jurassic calcarenites, Red Rock field, Webster Parish, Louisiana: Gulf Coast Assoc. Geol. Socs. Trans., v. 21, p. 125-138.

Boyd, D. R., 1963, Geology of the Golden Lane trend and related fields of the Tampico Embayment, in Geology of Peregrina Canyon and Sierra de El Abra, Mexico: Corpus Christi Geol. Soc. Ann. Field Trip Guidebook, p. 49-56.

Chimene, C. A., 1974, Arkansas' Walker Creek area should grow at both ends: Oil & Gas Journal, v. 72, p. 203-212.

Choquette, P. W., and J. D. Traut, 1963, Pennsylvanian carbonate reservoirs, Ismay field, Utah and Colorado, in Shelf carbonates of the Paradox basin: Four Corners Geol. Soc., p. 157-184.

Dunn, P. J., 1960, Harmattan-Elkton field, in Oil fields of Alberta: Alberta Soc. Petroleum Geol. Bull., v. 8, p. 128-129.

Edie, R. W., 1961, Devonian limestone reef reservoir, Swan Hills oil field, Alberta: Canadian Inst. Mining and Metall. Trans., v. 64, p. 278-285.

Elias, G. K., 1963, Habitat of Pennsylvanian algal bioherms, Four Corners area, in Shelf carbonates of the Paradox basin: Four Corners Geol. Soc., p. 185-203.

Ellis, G. D., 1962, Structures associated with the Albion-Scipio oil field trend: Michigan Geol. Survey Div., 86 p.

Enos, P., 1977, Tamabra Limestone of the Poza Rica trend, Cretaceous, Mexico, in H. E. Cook and P. Enos, eds., Deep-water carbonate environments: SEPM Spec. Pub. 25, p. 273-314.

Gill, D., 1977, The Belle River Mills gas field: productive Niagaran reefs encased by sabkha deposits, Michigan basin: Michigan Basin Geol. Soc. Spec. Paper 2, 187 p.

Guzman, E. J., 1955, Reef type stratigraphic traps in Mexico, in Geologia petrolera de Mexico: Asoc. Mexican Geologos Petroleros Bol., v. 7, p. 137-172.

———— 1967, Reef type stratigraphic traps in Mexico: 7th World Petroleum Cong. (Mexico), Proc., v. 2, p. 461-470.

Holden, R. N., 1967, Kawitt (Edwards) field, Karnes and DeWitt Counties, Texas, in Typical oil and gas fields of South Texas: Corpus Christi Geol. Soc., p. 91-95.

Hriskevich, M. E., 1967, Middle Devonian reefs of the Rainbow region of northwestern Canada, exploration and exploitation: 7th World Petroleum Cong. (Mexico), Proc., v. 3, p. 733-763.

Illing, L. V., G. V. Wood, and J. G. C. M. Fuller, 1967, Reservoir rocks and stratigraphic traps in non-reef carbonates: 7th World Petroleum Cong. (Mexico), Proc., v. 2, p. 487-499.

Kincheloe, J., et al, 1968, Greensburg pool, Green and Taylor Counties, in Geology and petroleum production of the Illinois basin: Illinois and Indiana-Kentucky Geol. Socs., p. 113-119.

Klovan, J. E., 1964, Facies analysis of the Redwater reef complex, Alberta, Canada: Bull. Canadian Petroleum Geol., v. 12, p. 1-100.

Kutney, E., 1966, Sylvan Lake field, in Oil fields of Alberta, supplement: Alberta Soc. Petroleum Geol. Bull., v. 14, p. 98-99.

Lauth, R. E., 1958, Desert Creek field, San Juan County, Utah, in Guidebook, geology of the Paradox basin: Intermtn. Assoc. Petroleum Geologists, p. 275-277.

Leavitt, E. M., 1968, Petrology, paleontology, Carson Creek North reef complex, Alberta: Bull. Canadian Petroleum Geol., v. 16, p. 298-413.

Linscott, R. O., 1958, Petrography and petrology of Ismay and Desert Creek zones, Four Corners region, in Guidebook, geology of of the Paradox basin: Intermtn. Assoc. Petroleum Geol., p. 146-152.

Lowenstam, H. A., 1950, Niagaran reefs of the Great Lakes area: Jour. Geology, v. 58, p. 430-487.

McCamis, J. G., and L. S. Griffith, 1967, Middle Devonian facies relationships, Zama area, Alberta: Bull. Canadian Petroleum Geol., v. 15, p. 434-467.

McNeely, W. M., 1960, Glenevis field, in Oil fields of Alberta: Alberta Soc. Petroleum Geol. Bull., v. 8, p. 116-117.

Mossop, G. D., 1972, Origin of the peripheral rim, Redwater reef, Alberta: Bull. Canadian Petroleum Geol., v. 20, p. 238-278.

Mountjoy, E. W., 1965, Stratigraphy of the Devonian Miette reef complex and associated strata, eastern Jasper National Park, Alberta: Canadian Geol. Survey Bull. 110, 113 p.

Murray, J. W., 1966, An oil-producing reef-fringed carbonate bank in the Upper Devonian Swan Hills Member, Judy Creek, Alberta: Bull. Canadian Petroleum Geol. v. 14, p. 1-103.

Myers, D. A., P. T. Stafford, and R. J. Burnside, 1956, Geology of the Late Paleozoic Horseshoe Atoll in west Texas: Bur. Econ. Geology, Univ. Texas, No. 5607, 113 p.

Nesbitt, J., 1958, Steelman field, southeastern Saskatchewan, in 2nd Williston basin symposium: Saskatchewan-North Dakota Geol. Socs., p. 94-99.

Oppel, T. W., 1960, Kenmitz field, in Oil and gas fields of southeastern New Mexico: Roswell Geol. Soc., p. 124-125.

Parker, E. R., 1956, Stratigraphy of the Charles Unit in the Midale field, Saskatchewan, in 1st Williston basin symposium: North Dakota-Saskatchewan Geol. Socs., p. 76-78.

Peterson, J. A., and H. R. Ohlen, 1963, Pennsylvanian shelf carbonates, Paradox basin, in Shelf carbonates of the Paradox basin: Four Corners Geol. Soc., p. 65-79.

Picard, M. D., 1958, Subsurface structure, Aneth and adjacent areas, San Juan County, Utah, in Guidebook, geology of the Paradox basin: Intermtn. Assoc. Petroleum Geol., p. 226-230.

Pray, L. C., and J. L. Wray, 1963, Porous algal facies (Pennsylvanian), Honaker Trail, San Juan Canyon, Utah, in Shelf carbonates of the Paradox basin: Four Corners Geol. Soc., p. 204-234.

Quigley, M. D., 1958, Aneth field and surrounding area, in Guidebook, geology, of the Paradox basin: Intermtn. Assoc. Petroleum Geol., p. 247-253.

Sigsby, R. J., 1976, Paleoenvironmental analysis of the Big Escambia Creek-Jay-Blackjack Creek field area: Gulf Coast Assoc. Geol. Soc. Trans., v. 26, p. 258-278.

Sloane, B. J., 1958, The subsurface Jurassic Bodcaw Sand in Louisiana: Louisiana Geol. Survey, Bull. 33, 33 p.

Stafford, P. T., 1954, Scurry field, in Occurrence of oil and gas in West Texas: Texas Bur. Econ. Geology, Publ. 5716, p. 295-302.

Terry, C. E., and J. J. Williams, 1969, The Idris "A" bioherm and oil field, Sirte basin, Libya—its commercial development, regional Paleocene geologic setting and stratigraphy, in The exploration for petroleum in Europe and North Africa: London, Inst. of Petroleum, p. 31-48.

Thomas, G. E., and H. S. Rhodes, 1961, Devonian limestone bank-atoll reservoirs of the Swan Hills area, Alberta: Alberta Soc. Petroleum Geol. Bull., v. 9, p. 29-38.

Vestal, J. H., 1950, Petroleum geology of the Smackover Formation of southern Arkansas: Arkansas Res. and Develop. Comm., Div. Geol. Inf. Circ. 14, p. 19.

Vogt, R. R., 1956, Alida field, southeastern Saskatchewan, in 1st Williston basin symposium: North Dakota-Saskatchewan Geol. Socs., p. 94-100.

Wakelyn, B. D., 1977, Petrology of the Smackover Formation (Jurassic): Perry and Stone Counties, Mississippi: Gulf Coast Assoc. Geol. Socs. Trans., v. 27, p. 386-408.

Zaffarano, R. F., et al, 1963, Reservoir oil characteristics, Greater Aneth area, Utah: U.S. Bur. Mines, Rept. Invest. 6196, 61 p.

Stratigraphic Traps in
Carbonate Rocks

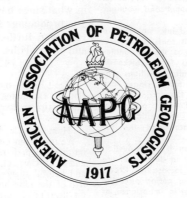

Geology of Beaverhill Lake Reefs, Swan Hills Area, Alberta[1]

C. R. HEMPHILL,[2] R. I. SMITH,[2] and F. SZABO[2]

Calgary, Alberta

Abstract The discovery in 1957 of oil in the remote Swan Hills region, 125 mi northwest of Edmonton, began a wave of exploration similar to that following the 1947 Leduc discovery which started the postwar oil boom in Western Canada. By the end of 1967 more than 1,800 wells had been drilled to explore and develop the Swan Hills region. Drilling has established in-place reserves of more than 5.9 billion bbl of oil and 4.5 trillion ft³ of gas.

Devonian sedimentary rocks unconformably overlie an eroded Cambrian section in the southeast part of the Swan Hills region; in the northwest part of the region, Devonian rocks lap onto the Precambrian granite of the Peace River arch.

Three positive features—the Tathlina uplift, Peace River arch, and the Western Alberta ridge—profoundly influenced Middle Devonian Upper Elk Point and Late Devonian Beaverhill Lake sedimentation. An embayment, shielded on the north by the emergent Peace River arch and on the south and west by the nearly emergent Western Alberta ridge, provided an environment conducive to reef development in the central Swan Hills region. Carbonate-bank deposition flanking the Western Alberta ridge in the south and southwestern part of the study area persisted throughout the time of Beaverhill Lake deposition. These beds merge with the overlying Woodbend reef system.

Recent changes proposed in Beaverhill Lake nomenclature include the elevation of the Beaverhill Lake to group status and the Swan Hills Member to formation status. The term Swan Hills Formation, as used herein, refers to the reef and carbonate-bank facies of the Beaverhill Lake Group, whereas the term Waterways Formation is applied to the offreef shale and limestone facies. The Swan Hills Formation is considered to be equivalent in age to the Calmut and younger members of the Waterways Formation.

The Swan Hills Formation is divided into Light Brown and Dark Brown members. Swan Hills reefs attained a thickness greater than 300 ft, whereas the carbonate-bank facies commonly exceeds 400 ft in thickness. Changing sedimentary and environmental conditions produced a complex reef facies; six major stages are postulated in the development of the undolomitized reef from which the Swan Hills field is producing. Stromatoporoids are the dominant reef-building organisms; abundant *Amphipora* characterize the restricted lagoonal facies.

Although the total impact of Swan Hills production on the provincial economy is difficult to determine, the $184 million paid by the industry to acquire Crown lands in the region during the 10-year period after the initial discovery attests to the economic importance of the Swan Hills producing region.

INTRODUCTION

Before 1957, oil production from the Swan Hills Formation of the Beaverhill Lake Group was unknown. Early in that year, the Virginia

Hills field was discovered and focused the attention of industry on a remote and heavily timbered area, approximately 125 mi northwest of Edmonton, Alberta (Fig. 1). Abundant rainfall, muskeg, lack of roads, and rugged topography (1,900–4,200 ft above sea level) made exploration difficult and costly.

The exposed stratigraphic sequence consists of the Late Cretaceous Edmonton Formation and early Tertiary Paskopoo Formation, capped with unconsolidated gravel similar to the Cypress Hills Conglomerate of southern Alberta and Saskatchewan (Russel, 1967).

The Swan Hills Beaverhill Lake reef discovery was made on a large farmout block by the Home Union H. B. Virginia Hills 9–20–65–13–W5M well. On January 31, 1957, a drill-stem test of a 30-ft interval in the Upper Devonian Beaverhill Lake section flowed 40° API oil to the surface. Within 30 days another successful Beaverhill Lake well (the Home et al. Regent Swan Hills 8–11–68–10–W5M) was drilled 25 mi northeast of the original discovery. This was the first well of the Swan Hills field. Paradoxically, the confirmation well for the Virginia Hills discovery was a dry hole, even though it was less than 1 mi away; and the 8–11–68–10–W5M well at Swan Hills found oil in what since has proved to be one of the poorest producing areas in the entire complex. However, further drilling around Virginia Hills and Swan Hills, and the impressive number of

[1] Read before the 53rd Annual Meeting of the Association, Oklahoma City, Oklahoma, April 24, 1968. Manuscript received, May 7, 1968; accepted, October 14, 1968.

[2] Home Oil Co. Ltd.

We thank our employer, Home Oil Co. Ltd., for permitting the preparation and publication of this paper, and J. L. Carr, chief geologist, for his encouragement and helpful suggestions. We are particularly appreciative of the long hours spent by W. C. Mackenzie and his drafting department in preparing the illustrations, and also thank W. Hriskevich who compiled the necessary statistical data. R. Sears was most helpful with the parts pertaining to reservoir data, as was A. B. Van Tine with the pressure-depth relations. Finally, we are grateful to Home Oil geologists, particularly H. H. Suter, for criticisms and helpful suggestions after reading the manuscript.

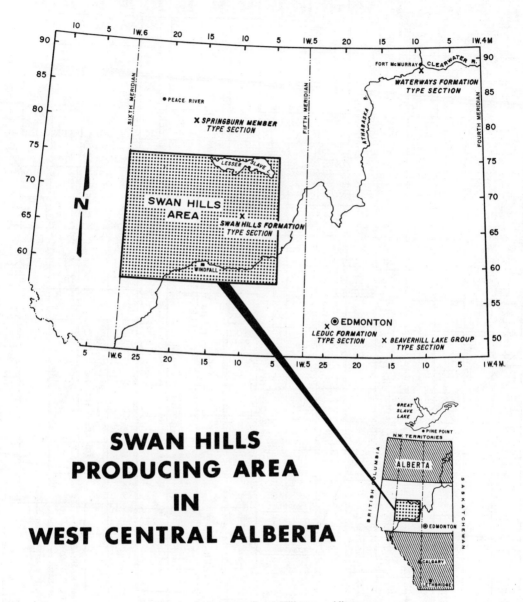

FIG. 1.—Index map, Swan Hills area, Alberta.

new-field discoveries assured continued activity through the region.

Within the map area few more than 12 wells had penetrated the Beaverhill Lake Group before the Home Oil discovery (Fig. 2). By the end of 1967, more than 1,800 wells had been drilled into or through rocks of the Beaverhill Lake Group, and resulted in the discovery of 12 oil fields and two gas fields in the Swan Hills Formation. Total initial in-place hydro-

carbons as of January 1, 1968, are estimated to have been in excess of 5.932 billion bbl and 4.512 trillion ft³ of gas (Oil and Gas Conservation Board, 1967).

The Swan Hills reef complex is an excellent example of an undolomitized Devonian carbonate bank-reef development. The availability of numerous cores has facilitated detailed facies studies on several fields. The main purpose of this paper, therefore, is to relate the entire pro-

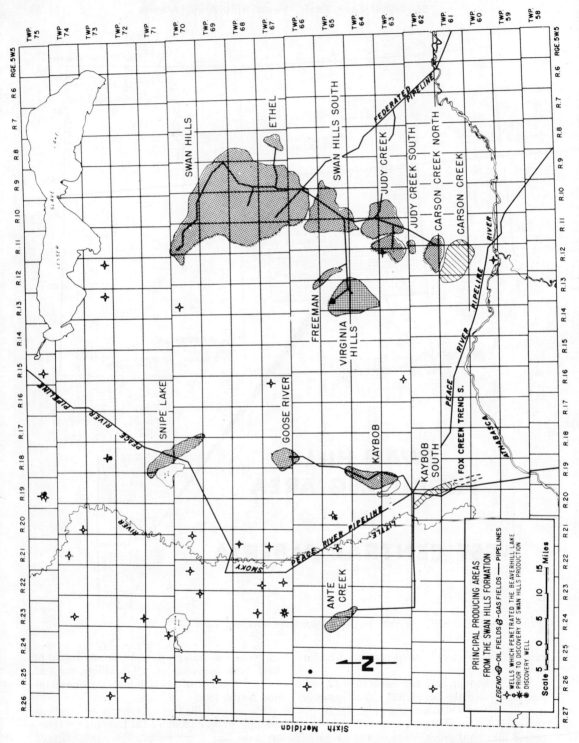

FIG. 2.—Location of Late Devonian Swan Hills Formation fields in Swan Hills region, Alberta. Well symbols are locations where rocks of Beaverhill Lake Group were penetrated before discovery. Gathering system and two major pipeline systems also are shown.

ductive area of the Swan Hills reef complexes to the regional, structural, and stratigraphic setting of the western Canadian sedimentary basin. A secondary purpose is to synthesize the published data on the Swan Hills and related beds and to propose some changes in the nomenclature to reduce confusion.

PREVIOUS WORK

Since Fong's (1959, 1960) original proposal of the type section and discussion of the geology of the Beaverhill Lake Formation, several other excellent studies have been made by Koch (1959), Carozzi (1961), and Thomas and Rhodes (1961). Edie (1961) was the first to make a detailed facies study of the Swan Hills field. Other outstanding studies include those by Fischbuch (1962), Brown (1963), Murray (1964, 1966), Jenik (1965), and Leavitt (1966).

MAP AREA

All present Swan Hills production is from a rectangular area in west-central Alberta, the dimensions of which are approximately 69 by 105 mi, or roughly 7,200 mi². The area is bounded on the east by 115° W long. and on the west by 118° W long. The south boundary is at 54° N lat and the north boundary is at 55° N lat. The area also can be described as lying between T60 and T71 and R8 and R25, W5M.

FIELDS

The Beaverhill Lake oil fields within the map area are, in order of decreasing importance, Swan Hills, Judy Creek, Swan Hills South, Virginia Hills, Kaybob, Carson Creek North, Snipe Lake, Goose River, Freeman, Ante Creek, Judy Creek South, and Ethel (Fig. 2). There are only two gas fields, Carson Creek and Kaybob South. Although most of the present fields are considered to be developed fully, some higher risk, marginal locations could be, and are being, drilled. In the Kaybob South field an important extension toward the southeast recently has been drilled at Fox Creek. Details are not available because of a highly competitive land situation.

DISCOVERY METHODS

Even though well control was extremely sparse for the deeper part of the section, several factors accounted for the gradual increase of exploration.

1. Decline in new discoveries of Leduc reefs (equivalent in age to the Woodbend) in the more accessible country, and the possibility of the presence in this area of other Leduc reef chains with large reserves.
2. Successful exploration along the Mississippian subcrop edges farther south and the possibility of similar conditions in this northern district.
3. An indicated thinning of the underlying Elk Point interval, possibly caused by the presence of a basement high which might have been the locus for reef development.
4. The numerous possibilities for the presence of Mesozoic sandstone bodies which were known to be productive in other parts of the province.
5. The presence of oölites, stromatoporoids, and *Amphipora* in the lower part of the Beaverhill Lake section in some of the older wells. (This indicated shallow-water shoaling and the potential for reef buildup.)
6. The large acreage blocks which could be assembled and the availability of additional offsetting Crown acreage through competitive bidding.
7. The discovery of oil in the "granite wash" just north of the Snipe Lake field.

Seismic studies have been of limited value in exploration for Beaverhill Lake reefs. A geophysical program was conducted on the acreage before the selection of the wellsite and, although the choice of drilling location was influenced by seismic information, the site selection in Virginia Hills was based on a somewhat nebulous feature. Similar seismic data were used to locate the first well in the Swan Hills field.

The value of seismic work in defining areas of Beaverhill Lake reefs has been argued strongly ever since. The arguments may be attributed to uncertainty in identification of the reflecting horizon, lack of velocity contrast between the reef and the enclosing rocks, and the absence of differential compaction or draping over the biohermal buildups. Both seismic structure and isochron maps have been used to select drillable locations, but generally the results have not permitted a good correlation of the well information with the seismic data.

GEOLOGIC HISTORY[3]

The western Canadian sedimentary basin is underlain by the westward continuation of Precambrian rocks of the Canadian shield. The oldest Cambrian basin on the shield rocks was restricted to the Cordilleran trough. It was a long, narrow marine trough in what is now northern British Columbia. The seaway straddled the British Columbia–Alberta boundary in the vicinity of the present Rocky Mountains.

[3] This discussion is mostly a synthesis of work by Van Hees and North (1964), Porter and Fuller (1964), Grayston *et al.* (1964), and Moyer *et al.* (1964).

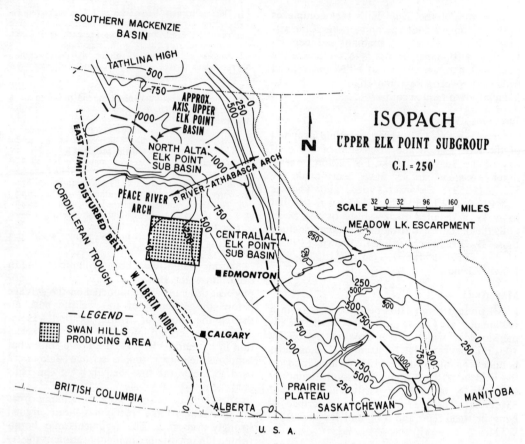

FIG. 3.—Isopach map, Upper Elk Point subgroup. Maximum thickness of Upper Elk Point strata is in northern Alberta and in Williston basin, Saskatchewan. Position of basinal axis is similar to position of Beaverhill Lake basinal axis shown in Fig. 4. Note emergent Western Alberta ridge and Peace River arch. CI = 250 ft. Redrawn from Grayston et al. (1964).

From this trough, Middle and Late Cambrian seas transgressed eastward over the cratonic shelf.

Sub-Devonian erosion removed Upper Cambrian rocks from all but the southeast part of the Swan Hills producing area. No Ordovician or Silurian rocks are present, presumably because of removal by pre-Devonian erosion. Basal Devonian strata unconformably overlie the eroded Cambrian section and in the northwest, lap onto Precambrian granite of the Peace River arch.

Thus, before the Devonian, Caledonian tectonism produced uplift and erosion which led to the widespread destruction of sedimentary beds. Three important positive areas—the Peace River–Athabasca arch, the Tathlina uplift, and the Western Alberta ridge—were present (Fig. 3), and were to exert an impor-

tant influence on sedimentation during Early, Middle, and part of Late Devonian times.

At the beginning of the Early Devonian, the Tathlina uplift and the Western Alberta ridge separated the moderately subsiding basin in Alberta from the MacKenzie basin on the north and the Cordilleran basin on the west. This interior basin, subdivided by the Peace River–Athabasca arch into northern and central Alberta Elk Point subbasins, received 600–1,300 ft of Lower Elk Point clastic and evaporite sediments. Topography controlled the lateral extent of the shallow Lower Elk Point sea.

Collapse of the eastern part of the Peace River–Athabasca arch near the end of Lower Elk Point deposition caused the northern and central Alberta subbasins to merge. The resulting basinal configuration remained essentially unaltered through the rest of the Elk Point de-

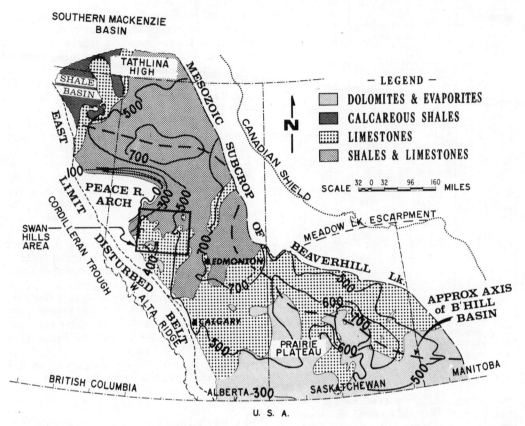

FIG. 4.—Isopach and lithofacies map, Late Devonian Beaverhill Lake Group. More than 700 ft of Beaverhill Lake present in northern and central Alberta and central Saskatchewan. Note progression of facies from outer marine shale in northeastern British Columbia to dolomite and evaporite in Southern Alberta, Saskatchewan, and Manitoba. Swan Hills producing area is southeast flank of Peace River arch. CI = 200 ft. Redrawn from Moyer (1964).

position, and for all of Beaverhill Lake deposition (Figs. 3, 4).

As subsidence and widespread incursion of more normal seawater continued, carbonate material was deposited early in Elk Point deposition. Keg River carbonate banks fringed the basin and patch reefs developed within the basin. For the first time during the Devonian the sea crossed the Meadow Lake escarpment, and sediments were deposited in southern Alberta, southern Saskatchewan, and southwest Manitoba.

The most rapid reef growth was in northeastern British Columbia; the resulting reef complex restricted the circulation of seawater into the Upper Elk Point basin. Reef growth terminated and evaporite accumulation predominated until near the end of Elk Point deposition in central Alberta. In northwestern Al-

berta fluctuating marine conditions prevailed generally near the end of Elk Point deposition. Crickmay (1957), Campbell (1950), and Law (1955) reported a hiatus between the Muskeg and Watt Mountain or Amco Formations in the Fort McMurray, Pine Point, and Steen River areas, respectively. Subaerial erosion and penecontemporaneous shallow-water deposition probably characterized the start of Watt Mountain deposition throughout the Upper Elk Point basin.

The sands of the Gilwood were derived from the Peace River arch and, according to Kramers and Lerbekmo (1967), represent ·a regressive-deltaic environment during Watt Mountain sedimentation. Other writers, including Guthrie (1956) and Fong (1960), believe the Gilwood Sandstone was deposited during shallow marine transgression. In the writers'

7

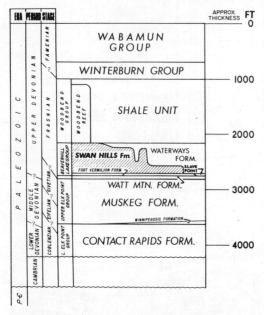

FIG. 5.—Devonian stratigraphic nomenclature, west-central Alberta. Chart shows position of Swan Hills Formation in geologic column. Relation of Swan Hills reef complex to enclosing Waterways Formation and younger Woodbend reefs is shown diagrammatically.

opinion, an oscillatory environment probably characterized Watt Mountain deposition and we agree with Griffin (1965), who wrote that the Fort Vermilion was deposited during final oscillatory conditions before the main transgressive phase which began during the time of Slave Point deposition.

The Late Devonian Beaverhill Lake transgression began with the deposition of Slave Point carbonate on a broad shelf in northeastern British Columbia, northern Alberta, and the adjacent part of the Northwest Territories. A carbonate reefoid-front facies, similar to the underlying Elk Point reefoid carbonate bank, developed in northeastern British Columbia.

Climatic changes and regional tectonism, accompanied by an influx of argillaceous material, destroyed the Slave Point biota (Griffin, 1965). Alternating limestone and shale characterize the rest of the Beaverhill Lake in northern Alberta and in the Edmonton area. However, in the Swan Hills region a shallow-water embayment, protected on the north by the emergent Peace River arch and flanked on the southwest by the Western Alberta ridge, provided a setting conducive to bank develop-

ment and subsequent reef growth. Emergence followed by deepening of water during late Beaverhill Lake deposition terminated Swan Hills reef growth. However, in places, notably the Windfall area, deposition of bank carbonate continued throughout Beaverhill Lake deposition. These carbonate beds grade into the overlying Woodbend reef. During the final deposition of the Beaverhill Lake, a fringing reef developed in the Springburn area on the south flank of the Peace River arch.

The overlying Woodbend consists of carbonate reefs and, in the offreef areas, green shale. The Winterburn consists of silty carbonate, red and green shale, and some anhydrite. The youngest Devonian unit, the Wabamun, is entirely carbonate.

The Devonian rocks within the study area and adjacent regions are subdivided in order of ascending age into Elk Point, Beaverhill Lake, Woodbend, Winterburn, and Wabamun Groups (Fig. 5). The total Devonian section along the eastern margin of the Swan Hills region is approximately 3,600 ft thick and thins markedly northwestward to 300 ft over the Peace River arch (Fig. 6). Within the east-central part of Alberta (i.e., over the central part of the Elk Point and Beaverhill Lake basins) the Devonian is thicker than 4,500 ft.

ELK POINT GROUP

McGehee (1949) first defined the Elk Point Formation. Belyea (1952) elevated it to group rank and Crickmay (1954) described the type section. The writers, following Grayston et al. (1964), subdivide the group into the Lower and Upper Elk Point subgroups, the exact ages of which are not settled. However, we tentatively follow Grayston et al. (1964), Basset (1961), and Hriskevich (1966), and place the Middle-Lower Devonian boundary at the Upper-Lower Elk Point subgroup division.

Opinion is divided concerning the boundary between the Elk Point and overlying Beaverhill Lake Group. The writers prefer to include the Watt Mountain Formation in the Beaverhill Lake Group. However, we recognize the fact that the controversy is far from resolved, hence do not assign the Watt Mountain Formation to either the Elk Point Group or Beaverhill Lake Group. Murray (1964) and Thomas and Rhodes (1961) placed the boundary at the top of the Fort Vermilion anhydrite. Fong (1960), Edie (1961), and Jenik (1965) considered the boundary to be at the top of the Watt Mountain Gilwood Sandstone Member. Grayston *et al.* (1964) believed that the Watt Mountain Formation belongs with the Beaverhill Lake depositional cycle and thus favored its inclusion with the Beaverhill Lake Group.

The Lower Elk Point subgroup consists mainly of terrigenous clastic and evaporite beds and ranges in thickness from 200 to 1,300 ft.

The Upper Elk Point subgroup is subdivided into the Winnipegosis Formation below and the Muskeg Formation above. The Winnipegosis is mainly carbonate, in contrast to the salt and anhydrite which predominate in the Muskeg Formation. Total thickness of this subgroup ranges from 400 ft in the eastern part of the Swan Hills region to more than 1,000 ft in the central part of the basin.

BEAVERHILL LAKE GROUP

The Beaverhill Lake section originally was defined by the Imperial Oil Geological Staff (1950). Common industry practice in the Swan Hills area has been to include basal carbonate beds of the Cooking Lake Formation with the Beaverhill Lake Group. The proper correlation is shown in Figure 7. The writers follow Leavitt (1966) and raise the Beaverhill Lake section to group rank. As used in this paper, the Beaverhill Lake Group includes, from base to top, the Fort Vermilion, Slave Point, Swan Hills, and Waterways Formations (Fig. 5).

The Beaverhill Lake type section in the Edmonton area was defined by Imperial Oil (1950) to include the interval from 4,325 to 5,047 ft (722 ft) in Anglo Canadian Beaverhill Lake (Lsd. 11, Sec. 11, T50, R17, W4M). Fong (1960) defined the Beaverhill Lake section in the Swan Hills producing area as the interval 8,020–8,543 ft (523 ft) in Home Regent "A" Swan Hills (Lsd. 10, Sec. 10, T67, R10, W5M). The top of the interval coincides with the highest beds of dark-brown limestone below bituminous shale. Subsequent drilling between the Swan Hills region and the Beaverhill Lake type section well has indicated that the upper boundary of the Beaverhill Lake as defined in the type section has been miscorrelated with the upper boundary established by usage in the Swan Hills region. This discrepancy is shown in Figure 7.

Beaverhill Lake strata are present through much of the western Canadian sedimentary basin, but are of maximum thickness in central Alberta, where drilling has established the presence of up to 750 ft of section (Fig. 4). Fig-

ure 4 shows that the axis of the Beaverhill Lake basin closely parallels that of the Upper Elk Point basin (Fig. 3). A marked thickness reduction of the Beaverhill Lake in northeastern British Columbia, though due partly to depositional thinning, has been interpreted by Griffin (1965) as primarily the result of post-Beaverhill Lake erosion.

Beaverhill Lake beds may be divided into five main facies: the outer shale facies, carbonate-front facies, inner alternating limestone and shale facies, inner reef facies, and shelf-margin carbonate-evaporate facies.

The carbonate-front facies, extending northeastward through northern British Columbia into the Northwest Territories, formed an effective barrier to the Beaverhill Lake shelf basin, profoundly influencing sedimentation there and on the shelf margin. Seaward, the carbonate-front facies is in abrupt contact with dark, deep-water marine shale, and the central part of the Beaverhill Lake basin contains the alternating limestone and shale facies of the Waterways Formation. Southward, the limestone and shale facies grades into shelf-margin carbonate and evaporite, limestone, and primary dolomite. In southern Alberta, Saskatchewan, and Manitoba, anhydrite is a major constituent of the carbonate-evaporite facies (Fig. 4).

The inner reef facies, designated the Swan Hills Formation, is restricted to an embayment south of the Peace River arch.

McLaren and Mountjoy (1962), on faunal evidence, correlate the Moberly and Mildred —upper members of the Waterways Formation —with the Flume Formation exposed in the front ranges of the Rocky Mountains south of the Athabasca River. Lower beds are correlative with the Flume Formation north of the Athabasca River.

Agreement is lacking on the age of the Beaverhill Lake Group. Mound (1966) established a Late Devonian age from conodont studies. Clark and Ethington (1965), on the basis of conodont data, placed all but the top few feet of the Flume Formation in the Middle Devonian. Norris (1963) assigned the Waterways Formation to the early Late Devonian and the Slave Point Formation in northeastern British Columbia to the Middle Devonian. McGill (1966), on the basis of ostracods, placed the Givetian-Frasnian boundary at the base of the Waterways Formation in the Lesser Slave Lake area in the northeastern corner of the Swan Hills region. Loranger (1965) placed the entire Beaverhill Lake and the lower part

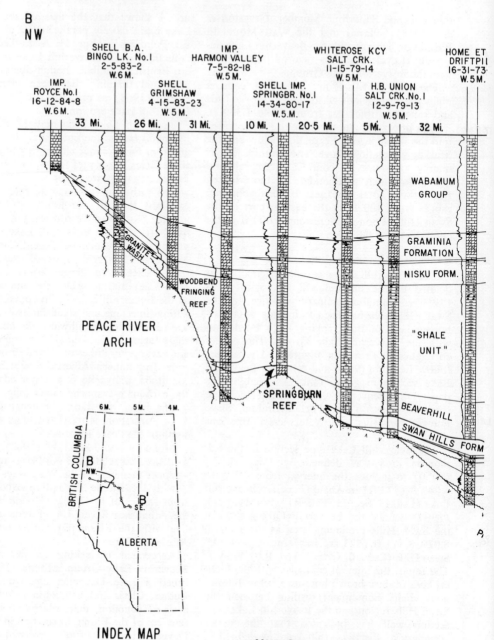

B
NW

IMP.
ROYCE No.1
16-12-84-8
W.6 M.

SHELL B.A.
BINGO LK. No.1
2-5-83-2
W.6 M.

SHELL
GRIMSHAW
4-15-83-23
W.5 M.

IMP.
HARMON VALLEY
7-5-82-18
W.5 M.

SHELL IMP.
SPRINGBR. No.1
14-34-80-17
W.5 M.

WHITEROSE KCY
SALT CRK.
11-15-79-14
W.5 M.

H.B. UNION
SALT CRK.No.1
12-9-79-13
W.5 M.

HOME ET
DRIFTP11
16-31-73
W.5 M.

33 Mi.   26 Mi.   31 Mi.   10 Mi.   20·5 Mi.   5 Mi.   32 Mi.

GRANITE WASH

WOODBEND
FRINGING
REEF

PEACE RIVER
ARCH

SPRINGBURN
REEF

WABAMUM
GROUP

GRAMINIA
FORMATION

NISKU FORM.

"SHALE
UNIT"

BEAVERHILL

SWAN HILLS FORM.

INDEX MAP

6 M.   5 M.   4 M.

BRITISH COLUMBIA

B
NW.

B'
SE.

ALBERTA

NW-SE
GENERALIZED DEVONIAN-CAMBRIAN
CROSS SECTION B-B'
OFF THE PEACE RIVER ARCH

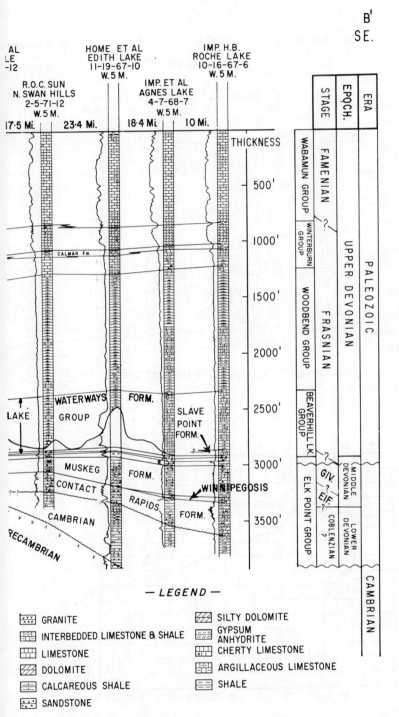

FIG. 6.—Northwest-southeast cross section **B-B′** showing thinning of Devonian over Peace River arch. Crest of arch was emergent until time of Wabamun deposition. Swan Hills Formation, deposited during early Late Devonian transgression, onlaps southeastern flank of Peace River arch. Vertical scale in feet.

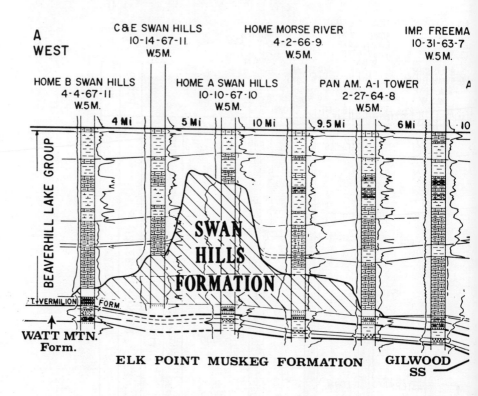

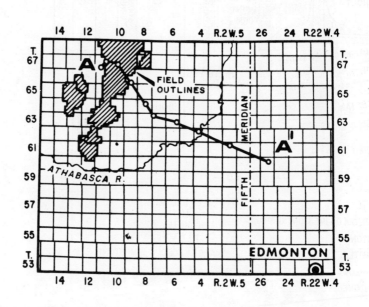

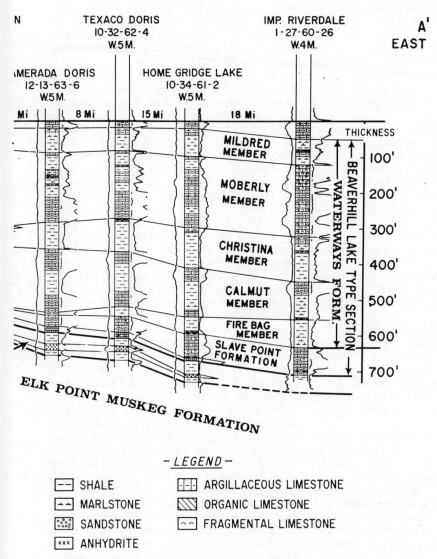

FIG. 7.—West to east cross section **A-A'** of Beaverhill Lake Group, through Swan Hills field, Alberta, into offreef basin area. Section demonstrates that Swan Hills Formation is time equivalent of Calmut and younger members of Waterways Formation, and not of Slave Point Formation as believed by some authors. Relation of upper Beaverhill Lake limestone in Swan Hills area to Beaverhill Lake type section should be noted. Thicknesses in feet.

of the overlying Woodbend Group in north-eastern Alberta in the Middle Devonian.

The writers assign a Late Devonian age to the Beaverhill Lake Group in the Swan Hills region. This does not conflict with the findings of Norris, because the Slave Point Formation is the product of initial deposition of an advancing sea from the north, and thus as a transgressive deposit should become progressively younger southward.

Fort Vermilion Formation.—The name Fort Vermilion Member was introduced by Law (1955) for an evaporite unit at the base of the Slave Point Formation, which overlies the Watt Mountain Formation in the subsurface of northwestern Alberta. This anhydrite unit is only 23 ft thick in the California Standard Steen River type well (Lsd. 2, Sec. 22, T117, R5, W6M) but thickens eastward to 120 ft at the Hudson's Bay No. 1 Fort Vermilion (Sec. 32, T104, R8, W5M). Fort Vermilion anhydrite is widespread; subsequent drilling has established Fong's (1960) "Basal Beaverhill Lake" anhydrite unit as correlative with the Fort Vermilion.

Norris (1963) raised the Fort Vermilion Member to formation status. The writers concur with Norris, because the Fort Vermilion beds are widespread and because we agree with Griffin (1965), that the Fort Vermilion was deposited during final oscillatory conditions before the main Slave Point transgressive phase began.

Slave Point Formation.—Cameron (1918) assigned the name Slave Point Formation to scattered exposures of thin-bedded, medium-grained, dark-gray bituminous limestone, which he concluded were of Middle Devonian age. The limestone crops out on the north and south shores of Great Slave Lake. Cameron originally estimated the total section to be 160 ft thick, but in 1922 revised this estimate to 200 ft.

Campbell (1950) correlated a 310-ft section of limestone, shale, and dolomite penetrated in bore holes at Pine Point on the south shore of Great Slave Lake with Cameron's Slave Point section. Campbell's subsurface section includes 170 ft of fine-grained, stromatoporoidal limestone overlying 11 ft of dark-greenish-gray shale, termed the Amco Formation, which unconformably overlies 129 ft of fossiliferous limestone and dolomitic limestone. Law (1955) noted a hiatus between the Watt Mountain and Muskeg Formations penetrated in the California Standard Steen River (2–22–117–5–W6M),

in northwest Alberta, and he correlated the Watt Mountain with the Amco Formation. Law concluded that only the upper 170 ft of the carbonate rocks in the Pine Point area subsurface correlate with the Slave Point Formation of the plains (informal usage).

The Slave Point Formation of the plains attains a maximum thickness of 500 ft, in northwestern Alberta, north of the Peace River arch, northeastern British Columbia, and the District of MacKenzie. In northeastern British Columbia and the District of MacKenzie, there is an abrupt change in facies between the Slave Point carbonate facies front and the basinal Otter Park Shale. Gray and Kassube (1963) describe the facies front as being rich in stromatoporoids and *Amphipora*. Griffin (1965) defined five rock types in the vicinity of the facies front: dark stromatoporoid-bearing calcilutite, stromatoporoid biosparite, stromatoporoid biomicrite, light-colored micrite (and fossil micrite), and white and gray dolomite. Dolomitization has destroyed many of the original lithic characteristics, but the abundance of stromatoporoids suggests that the front facies is in fact a reef chain.

The Slave Point limestone thins eastward, and also south of the Peace River arch. Crickmay (1957) noted 5.5 ft of magnesium limestone between the Waterways and Muskeg Formations at the Bear Biltmore 7–11–87–17–W4M well near Fort McMurray. On the basis of fossil content, Crickmay considered this unit to be correlative with the Slave Point Formation of the outcrop area. Subsequently, Belyea (1952) correlated the unit with the basal limestone member of the Beaverhill Lake Formation type section (*i.e.*, she considered the Beaverhill Lake section in the Edmonton area to be equivalent to the Waterways Formation plus the 5.5-ft magnesium limestone section at the Waterways type locality). Norris (1963) disputed Crickmay's correlation of these beds with the Slave Point Formation and proposed the term "Livock River Formation" to include this section. Griffin (1965) confirmed Crickmay's interpretation, however, by establishing, on the basis of mechanical-log and lithologic analysis, a convincing correlation between Norris's Livock River Formation and the Slave Point Formation in northeastern British Columbia. Thus, the basal 35-ft limestone unit in the type Beaverhill Lake Formation correlates with the Slave Point Formation.

Swan Hills Formation.—Fong (1960) applied the name Swan Hills Member to the productive unit of the Beaverhill Lake in the Swan Hills area. The type section is the interval at 8,167–8,500 ft (333 ft) in the Home Regent "A" Swan Hills 10–10–67–10–W5M well. Murray (1966) proposed that the Swan Hills Member be raised to formation rank, a view also held by Leavitt (1966). The writers concur. Drilling since 1960 has established the presence of Swan Hills rocks across an area of approximately 7,000 mi². Its widespread occurrence and the economic importance support elevation to formation rank.

Fong defined the Swan Hills as an organic bioclastic limestone unit overlying and gradational with anhydrite and shale now known to be correlative with the Fort Vermilion Formation. The Swan Hills Formation is divisible into a lower Dark Brown member and an upper Light Brown member. In a typical section (Fig. 8) the Dark Brown member is 110 ft thick but may range from 80 to 160 ft, and consists of dark-brown calcarenite and calcilutite with some reef-building organisms. The Dark Brown member grades upward into the Light Brown member, which has a maximum thickness of 320 ft, and consists of calcarenite, calcilutite, and biogenic carbonate.

The Swan Hills Formation consists of organic carbonate deposits. The central part of the Swan Hills region contains organic reefs as thick as, or slightly thicker than, 300 ft. The carbonate-bank deposits which characterize the southwestern part of the region are more than 400 ft thick and grade upward into the overlying Woodbend reef deposits.

The Swan Hills is transgressive, becoming younger northwestward in the direction of the Peace River arch (Fig. 6). The unit wedges out against the lower flank of the Peace River arch. An upper Beaverhill Lake reef termed the Springburn Member is interpreted to be a local fringe reef. The Springburn is a forerunner to the extensive Woodbend fringe reef and is considered by the writers to be separate from the Swan Hills Formation.

Murray (1966) observed a very sharp contact between the reef and offreef facies in the Judy Creek area and suggested that most of the basin facies is younger than the Swan Hills Formation. Leavitt (1966) found in the Carson Creek area, in addition to a sharp reef-offreef boundary, a deeper water fauna in the adjacent offreef facies and concluded that there was considerable relief between the reef complex and the offreef basin. Apparently, this relation is not characteristic of the entire Swan Hills re-

gion, for Fischbuch (1962) reported an interfingering of Swan Hills carbonate rocks with offreef strata near the south end of the Kaybob field.

Waterways Formation.—The name Waterways originally was suggested by Warren (1933) for the limestone and shale which crop out at the confluence of the Clearwater and Athabasca Rivers in northeastern Alberta. Warren's Waterways section, as determined from outcrop and nearby salt wells, totals 405 ft. On the basis of the section penetrated in the Bear Biltmore 7–11 well, Crickmay (1957) also included beds removed by erosion in the Fort McMurray area. The type subsurface section penetrated at Bear Biltmore (Lsd. 7, Sec. 11, T87, R17, W4M), is 740 ft and was subdivided by Crickmay (1957) into five units designated in ascending order the Firebag, Calmut, Christina, Moberly, and Mildred Lake Members.

The Waterways Formation, consisting of alternating brown fragmental limestone and greenish-gray shale, is separated from the underlying Elk Point Group by the 5.5-ft Slave Point magnesian limestone bed. Griffin (1965) traced Crickmay's (1957) five-member subdivision of the Waterways Formation as far north as California Standard Mikkwa (Lsd. 12, Sec. 23, T98, R21, W4M). In addition, Griffin's correlation of the top of the Calmut Member in northeastern British Columbia provided convincing support for his suggestion that thinning of the shale-limestone sequence in northwesternmost Alberta and northeastern British Columbia is due to (1) facies equivalence of the Slave Point Formation (carbonate) with the Waterways Formation (limestone and shale) and (2) progressive northwestward erosional truncation of the Beaverhill Lake Group. These relations are illustrated in Figure 9.

AGE RELATIONS OF SWAN HILLS, SLAVE POINT, AND WATERWAYS FORMATIONS

Warren (1957) observed that ". . . the oil-bearing horizon within the Beaverhill Lake Formation [*i.e.,* the Swan Hills Formation] south of Lesser Slave Lake carries a fauna younger than that of the Firebag Member of the Waterways Formation [Fig. 7] and thus is younger than the Slave Point Formation."

According to Crickmay (1957), *Atrypa* aff. *A. independensis* is a guide fossil everywhere for the Slave Point. Koch (1959) noted the abundance of this species within the basal part of the Swan Hills Formation. Its presence sug-

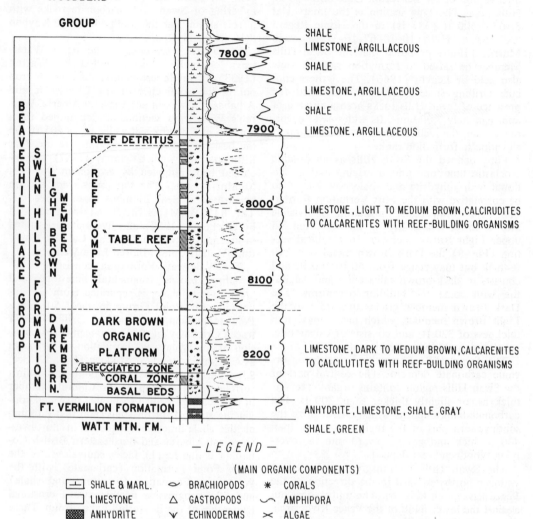

FIG. 8.—Typical Beaverhill Lake section in Swan Hills field, Alberta. Well is Home Regent Swan Hills 4-28-67-10-W5M. KB elev. = 3,322.1 ft. Well was cored almost completely through Swan Hills Formation, permitting excellent control for facies divisions shown. Depths in feet below KB.

gests that the lower beds of this formation correlate with the Slave Point Formation. Koch considered, however, the tendency for *Atrypa* of the *A. independensis* type to migrate with the nearshore carbonate facies (*i.e.*, it is a facies-controlled organism), and accordingly he preferred to correlate the basal Swan Hills with the Firebag Member.

The writers concur with Warren and believe that the basal part of the Swan Hills Formation in the Swan Hills field correlates with the Calmut Member of the Waterways Formation. Moreover, the Slave Point Formation is known to wedge out east of the Swan Hills fields and thus was not deposited within the designated Swan Hills region. These relations are illustrated in Figure 7.

Woodbend Group.—The contact between the Beaverhill Lake and Woodbend Groups is gradational. Bioclastic limestone and shale of

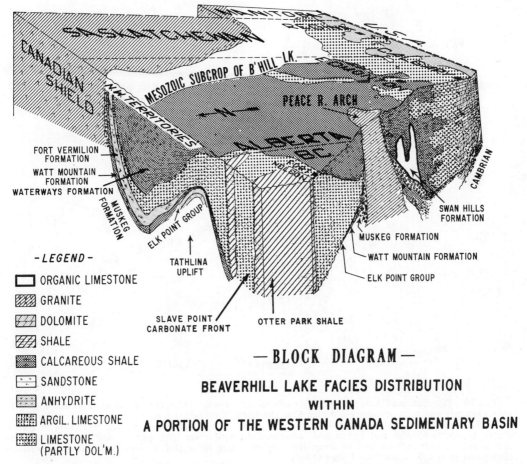

- LEGEND -

☐ ORGANIC LIMESTONE

▨ GRANITE

▨ DOLOMITE

▨ SHALE

▨ CALCAREOUS SHALE

▨ SANDSTONE

▨ ANHYDRITE

▨ ARGIL. LIMESTONE

▨ LIMESTONE
   (PARTLY DOL'M.)

— BLOCK DIAGRAM —

BEAVERHILL LAKE FACIES DISTRIBUTION
WITHIN
A PORTION OF THE WESTERN CANADA SEDIMENTARY BASIN

FIG. 9.—Block diagram showing Beaverhill Lake facies distribution, and position of Swan Hills Formation with respect to Peace River arch and to various Beaverhill Lake facies. View is toward southeast. Basinal Otter Park shale facies is in abrupt contact with Slave Point carbonate-front facies. Calcareous shale and argillaceous limestone of Waterways Formation occupy much of Alberta and grade southeastward into primary dolomite, anhydrite, and limestone facies.

the Cooking Lake Formation compose the basal part of the Woodbend Group. The Cooking Lake Formation formed a platform for subsequent Leduc reef development in central Alberta (Fig. 10) where, in the offreef areas, it is overlain by the Duvernay bituminous shale and interbeds of dark-brown limestone. The Duvernay Formation is overlain by the Ireton Formation, a green shale section flanking and overlying the Leduc reefs.

In the Swan Hills region, the Duvernay-Cooking Lake section is represented principally by greenish-gray calcareous shale; accordingly, the Ireton-Duvernay-Cooking Lake section is referred to herein as the "shale unit." Unlike

Fong's (1960) "shale unit," it does not include the Nisku Formation of the Winterburn Group. Thus, the section in this paper is identical with the "shale unit" of Murray (1966).

Younger Devonian.—The Woodbend is overlain by the Winterburn and Wabamun Groups. The Woodbend includes shale, siltstone, anhydrite, dolomite, and silty dolomite. The Wabamun consists mostly of dolomitic limestone and some anhydrite.

STRATIGRAPHY OF SWAN HILLS FORMATION

The Swan Hills Formation is divided into two members on the basis of color and morphologic characteristics—a lower or Dark

17

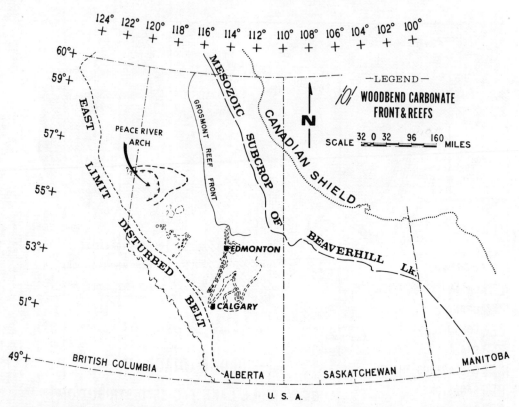

FIG. 10.—Distribution of Late Devonian Woodbend reef systems. Devonian carbonate front and reefing in Alberta reached maximum development during Late Devonian (Woodbend) time. Grosmont reef complex under-lies northeastern Alberta and is followed in clockwise direction by Leduc-Meadowbrook reef chain in central Alberta. Simonette-Windfall reef system and Sturgeon Lake atoll cover large part of west-central Alberta in-cluding western and southern part of Swan Hills region. Peace River arch is flanked by broad, continuous, fringing reef system. Redrawn from Belyea (1964).

Brown member and an upper or Light Brown member (Fig. 8).

The Dark Brown member is a widespread organic platform or carbonate bank, covering an extensive area of west-central Alberta. It is fringed on the northeast, north, and northwest by stromatoporoidal limestone, which is inter-preted as an organic reef (Fig. 11). This mem-ber represents the first transgressive phase of the Late Devonian sea.

Generally there is an abrupt facies change toward the east and north between the Dark Brown member and the calcareous shale and argillaceous limestone of the Waterways For-mation. On the west the entire Swan Hills For-mation is terminated by onlap on the Peace River arch.

The thickness of the Dark Brown member ranges from 80 ft in the Deer Mountain–

House Mountain area to approximately 160 ft in the vicinity of Carson Creek (Fig. 12). Thin-section studies show that the dark color is caused mainly by the presence of pyrobitumen and residual oil in the fine matrix rather than by a high argillaceous content.

Frame-building organisms became estab-lished at numerous localities on the top of the Dark Brown member and formed a bioherm or "reef complex," as used in this paper. This complex probably grew on topographic highs or positive areas of the underlying carbonate bank, hence its lateral extent is more restricted than that of the platform on which it rests.

The Swan Hills, Judy Creek, Carson Creek North, Kaybob, and Goose River fields are sep-arate reeflike buildups on the platform, sepa-rated from each other by surge channels and the associated offreef shale and limestone beds

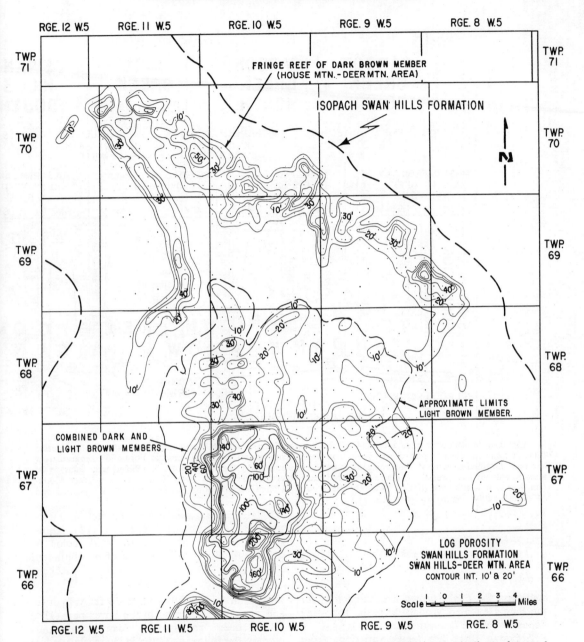

FIG. 11.—Isoporosity map of Swan Hills Formation, based on logs. Map emphasizes productive reef-run and main buildup area. Patch reefs and reef outwash are loci of porosity in intervening area. Porous trend along northern edge originally was called House Mountain–Deer Mountain area but now is included in area designated as Swan Hills field. Note line showing limits of Light Brown member. CI = 10 and 20 ft.

of the Waterways Formation (Figs. 13–15).

South and southwest of the Swan Hills region, the Light Brown member formed a massive carbonate bank with reef-building organisms around the rim of the widespread buildup. The Carson Creek, Virginia Hills, and Ante Creek fields are associated with this reef-rimmed bank (Fig. 14). Farther south and

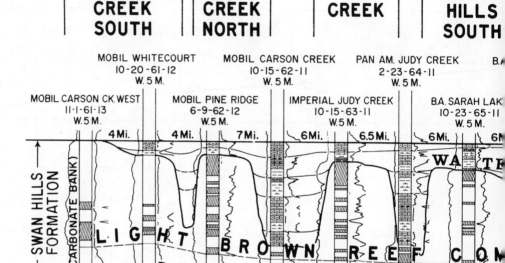

FIG. 12.—South-north cross section **C-C'** of Beaverhill Lake Group, north-central Alberta. Section passes through fields of eastern reef chain. Shale and argillaceous limestone of Waterways Formation separate reef buildups. Gradual southward thickening of Dark Brown and Light Brown members, combined with thinning of gross Beaverhill Lake, results in entire Beaverhill Lake interval consisting of Swan Hills carbonates. Deer Mountain field now is northern part of Swan Hill field. Area through which section passes is shown on Figures 2 and 14; exact location shown on Figure 21.

southwest the bank formed the foundation for the Late Devonian Woodbend reefs, following deposition of the Beaverhill Lake.

Although there are facies variations from field to field in the area, several facies types predominate in all fields (Fig. 8).

Dark Brown Member

"Basal beds."—The lowermost bed of the "Basal beds" overlies the Fort Vermilion anhydrite and shale, and consists of an argillaceous, cryptograined, dark-brown to black limestone

ranging in thickness from 3 to 14 ft. Anhydrite and pyrite are common in the lower part. The "Basal beds" are sparsely fossiliferous, and contain the following fauna in decreasing order of predominance: brachiopods (*Atrypa*), ostracods, echinoderms, and crinoids.

The uppermost units of the "Basal beds" are transitional with the overlying "Coral zone," and have similar organic content. These beds were deposited in aerated seawater of normal salinity. The presence of reef-building organisms in the upper part, the higher percentage

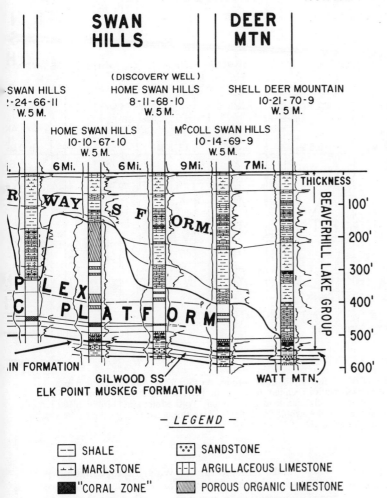

C'
NORTH

SWAN HILLS   DEER MTN

(DISCOVERY WELL)

SWAN HILLS   HOME SWAN HILLS   SHELL DEER MOUNTAIN
:-24-66-11   8-11-68-10   10-21-70-9
W. 5 M.   W. 5 M.   W. 5 M.

HOME SWAN HILLS   MᶜCOLL SWAN HILLS
10-10-67-10   10-14-69-9
W. 5 M.   W. 5 M.

6 Mi.   6 Mi.   9 Mi.   7 Mi.

THICKNESS

100'
200'
300'
400'
500'
600'

BEAVERHILL LAKE GROUP

GILWOOD SS   WATT MTN.
ELK POINT MUSKEG FORMATION

- LEGEND -

SHALE   SANDSTONE
MARLSTONE   ARGILLACEOUS LIMESTONE
"CORAL ZONE"   POROUS ORGANIC LIMESTONE
ANHYDRITE   FRAGMENTAL LIMESTONE
                   DENSE ORGANIC LIMESTONE

of skeletal material, and the local presence of sparry calcite cement suggest deposition in shallower water than the adjacent offreef basal beds of the Waterways Formation on the east (Fig. 7).

"Coral zone."—Overlying the "Basal beds" is light-brown, fossiliferous limestone of great lateral extent, called the "Coral zone" by Fong (1959). The thickness ranges from 4 to 20 ft. The lower and upper contacts are transitional. This *in situ* reefal biofacies contains the lowest zone of porosity in the productive areas. The thickness, porosity, and permeability of the "Coral zone" improve where the Light Brown member also is well developed. In the marginal areas of the dark-brown organic platform this facies is not recognizable, but is replaced by stromatoporoid limestone. Organic components of the "Coral zone" in decreasing order of predominance are *Thamnopora* corals, massive and tabular stromatoporoids, *Amphipora,* and *Stachyoides.* Skeletal fragments and grains, and mud-supported carbonate rocks (calcilutite) are the main matrix components.

21

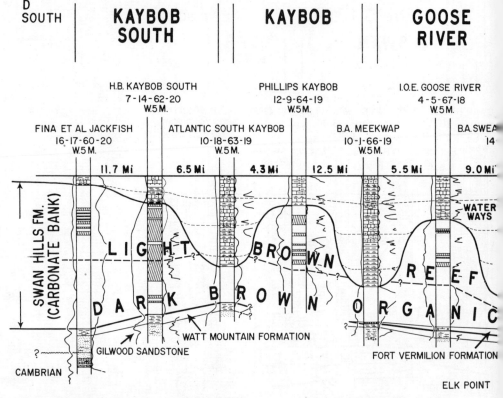

FIG. 13.—South-north cross-section **D-D'** of Beaverhill Lake Group, north-central Alberta. Section passes through western Swan Hills fields and shows relations similar to those in Fig. 12. Gradual pinchout of Fort Vermilion Formation causes basal beds of Beaverhill Lake to overlie directly Watt Mountain Formation. Gross Beaverhill Lake thins markedly between Kaybob and Kaybob South fields. Area through which sections passes is shown on Figures 14 and 21; exact location on Figure 21.

Above the "Coral zone" are dense beds which completely separate it from the Light Brown member. As a result, the "Coral zone" in the Swan Hills field is a separate producing zone of the Swan Hills Formation.

Stromatoporoid reef front.—Around the northern and northeastern margins of the Dark Brown member platform high water energy fostered the growth of a reef body with a rigid framework. The slightly different organic composition in the different reefs probably is related to local energy conditions and food supply. Along the eastern edge of the Swan Hills field, stromatoporoids predominate, but the increasing numbers of *Amphipora* and the greater amount of carbonate-mud matrix suggest a lower energy environment. Toward the central part of the platform, the reef front interfingers with the dark brown *Amphipora*

beds, and the offreef Waterways Formation is discordant. Organic components, in order of decreasing importance, are tabular and massive stromatoporoids, algae, brachiopods, cup corals, *Amphipora,* and crinoids. The matrix ranges from coarse reef detritus to fine carbonate mud and in places consists of skeletal grains, intraclasts, and some sparry calcite.

The porosity common to these beds is associated with both the organic framework and matrix. Hydrocarbon production is from the stromatoporoidal reef front at the edge of the platform north and northeast of the Swan Hills and Virginia Hills fields. Similar beds also are present in the marginal areas of the Light Brown member in the upper part of the reef complex and, where such beds are developed, they also are productive.

"Brecciated zone."—This term applies to a

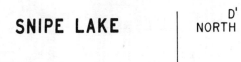

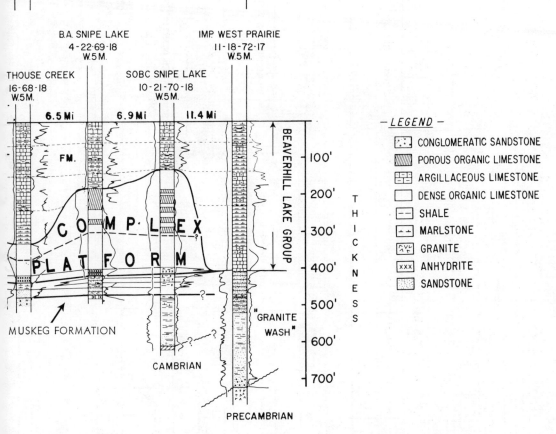

SNIPE LAKE   D'
                                NORTH

B.A. SNIPE LAKE
4-22-69-18
W.5 M.

IMP. WEST PRAIRIE
11-18-72-17
W.5 M.

THOUSE CREEK
16-68-18
W.5 M.

SOBC SNIPE LAKE
10-21-70-18
W.5 M.

6.5 Mi   6.9 Mi   11.4 Mi

BEAVERHILL LAKE GROUP

FM.

COMPLEX
PLATFORM

MUSKEG FORMATION

"GRANITE
WASH"

CAMBRIAN

PRECAMBRIAN

THICKNESS

100'
200'
300'
400'
500'
600'
700'

— LEGEND —

CONGLOMERATIC SANDSTONE
POROUS ORGANIC LIMESTONE
ARGILLACEOUS LIMESTONE
DENSE ORGANIC LIMESTONE
SHALE
MARLSTONE
GRANITE
ANHYDRITE
SANDSTONE

stromatoporoid bed overlying the "Coral zone" that has been observed in most of the Swan Hills–Ante Creek area. The name refers to the brecciated appearance of the cores, which contain numerous circular and semicircular, partly broken stromatoporoid colonies. However, examination of the cores proved that the appearance is caused by the growth pattern of the stromatoporoids. Only very limited mechanical transport is indicated, and the zone is not truly "brecciated."

The "Brecciated zone" interfingers with the stromatoporoidal reef front on the north and northeast, and with the *Amphipora* beds of the Dark Brown member on the west and southwest. The thickness ranges from 0 to 50 ft. Organic components in order of decreasing abundance include bulbous stromatoporoids, *Amphipora,* and corals. The matrix consists of fine

skeletal grains, intraclasts, and carbonate mud. Pyrobitumen also is common. The organisms of the "Brecciated zone" apparently accumulated in a lower energy environment than those of the stromatoporoidal reef front—probably in a restricted shelf lagoon. Porosity is sparingly present in association with stromatoporoids.

Dark brown Amphipora *beds.*—The dark-brown *Amphipora* beds extend throughout the Swan Hills region, and form the most common unit of the Dark Brown organic platform. The thickness ranges from 40 to 100 ft, generally increasing southward. There also are local variations in the thickness over areas where the Light Brown member is developed. The base of the unit is at the top of the "Coral zone," or at the top of the "Brecciated zone" where the latter is present. The unit interfingers with the stromatoporoidal reef front facies on the east

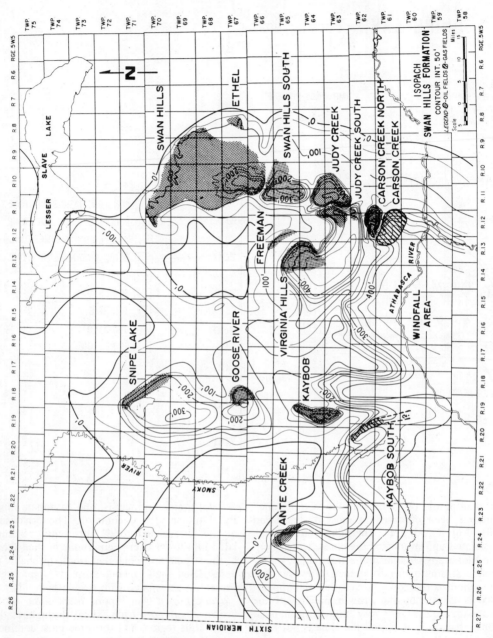

Fig. 14.—Isopach map of Swan Hills Formation. CI = 50 ft.

and northeast. The *Amphipora* content ranges from 0 to 60 percent, and the unit generally is less fossiliferous near the base. A few bulbous stromatoporoids also are present. Organic components, in order of decreasing importance, are *Amphipora,* bulbous stromatoporoids, brachiopods, and gastropods. Laminated limestone beds

with sparry calcite also are common and generally are devoid of organisms. The matrix ranges from poorly sorted silt- and sand-size grains and pellets to carbonate mud, or calcilutite. Skeletal grains also are recognizable.

Porosity is developed only locally in these beds, and is associated with some *Amphipora-*

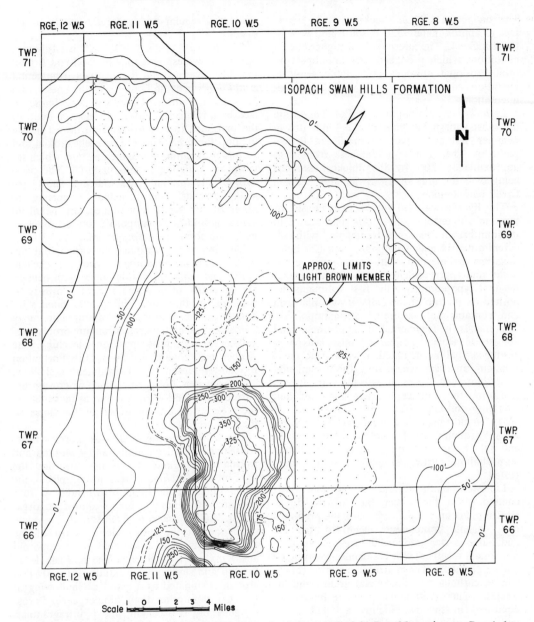

Fig. 15.—Detailed isopach map of Swan Hills Formation in Swan Hills field–Deer Mountain area. Broad platform is present between edge and Swan Hills main buildup. Rigid wave-resistant stromatoporoid reef wall first developed on northeast edge of platform.

rich zones. A little void space is present in a few places in granular or pelleted limestone, but for all practical purposes the dark-brown *Amphipora* beds do not contribute significantly to the pore volume of the Swan Hills Formation. This unit reflects sedimentary and environmental conditions of a restricted shelf lagoon, and the laminated limestone indicates occasional supratidal-flat conditions.

Light Brown Member

The thickness of the Light Brown member changes more abruptly than that of the gently sloping Dark Brown platform on which it lies.

The member ranges in thickness from a few feet to approximately 320 ft south and west of Carson Creek. The alternation of higher energy conditions which produced reef buildups with brief erosional episodes and stable sea-level conditions caused considerable variation in sedimentation.

Light-brown Amphipora *beds.*—These are the most widespread beds of the Light Brown member and cover part of the Dark Brown platform. They range in thickness from 0 to approximately 200 ft, the thickest sections being over the area of the Swan Hills–Carson Creek reef complex.

The light-brown *Amphipora* beds consist of two main limestone types: (1) light-brown to medium-brown limestone, fragmental, with calcarenite matrix and *Amphipora;* and (2) light-olive-brown limestone, with *Amphipora,* calcilutite, and sparry calcite matrix. Limestone of the first type comprises the first sediments deposited on the top of the Dark Brown platform and contains only scattered stromatoporoids. The matrix is fragmental in appearance and consists of skeletal grains and intraclasts, suggesting a shallow, agitated, open-marine environment. In areas where only this fragmental *Amphipora* bed is present over the Dark Brown platform, the contact between the reef and overlying offreef Waterways Formation is sharp. In most places the organisms of the Swan Hills Formation are truncated at the contact, and only a thin zone of pyrite is present; thus very shallow postdepositional erosion is suggested.

Limestone of the second type is found most commonly in the center part of the Light Brown member, *Amphipora* being the predominant organic component. Dendroid, tabular, and bulbous stromatoporoids are present in a few areas, together with ostracods, gastropods, and calcispheres. The matrix consists of sparry calcite, carbonate mud, skeletal grains, and intraclasts. These sediments appear to have been deposited in shallow, slightly agitated water with low turbidity, typical of a restricted shelf-lagoon environment.

Porosity development commonly is associated with the matrix and organisms of the light-brown fragmental *Amphipora* beds, whereas in the light-olive-brown limestone porosity is irregular and generally poorly developed.

"Table reef" (Edie, 1961).—This easily recognizable, widespread zone of the Light Brown member overlies the light-brown fragmental *Amphipora* beds and ranges in thickness from 15 to 25 ft. Its broad lateral extent is characterized by a relatively uniform organic content. In the marginal areas north and northwest of the buildup the organisms are severely broken and reworked, forming a fragmental zone which interfingers with the *in situ* material. The presence of the broken rocks indicates short erosional periods. The principal organisms, in order of decreasing abundance, are dendroid stomatoporoids, Solenoporoid algae, brachiopods, cup corals, and *Amphipora.*

The matrix is considerably varied, but the most common constituents are the skeletal grains and debris of various sizes, with negligible amounts of carbonate mud. Porosity is common in this zone and is associated with both organisms and matrix. Interorganic vuggy porosity present between stromatoporoids suggests secondary, postdepositional leaching. The "Table reef" zone of the Swan Hills Formation probably represents an environment of shallow, agitated water. Local emergence or erosion occurred at the end of deposition, as indicated by the presence of organic debris on the slopes of the reef.

Porous calcarenite beds.—A porous zone, consisting of skeletal grains and reworked and transported organic fragments, overlies the light-brown *Amphipora* beds and, in places, the upper slopes of the Light Brown buildup. In a few areas this zone contains almost no organisms, and consists of well-sorted carbonate sandstone; in the upper slopes, most of the zone consists of well-rounded stromatoporoid and *Amphipora* fragments. The reef rubble on the top and slopes of the buildup suggests that growth terminated as a result of shallowing of the seawater rather than submergence. Porosity generally is excellent, and is intragranular. Intraorganic void space also is common, decreasing slightly toward the central part of the buildup. This unit is one of the most important reservoir rocks of the area. The porous calcarenite beds probably were formed in the most exposed parts of the organic buildup, in shal-

---

Fig. 16.—South-north sections showing stages I–VI in development of Swan Hills reefing, as interpreted for Swan Hills field. Similar stages can be postulated for other Beaverhill Lake fields.

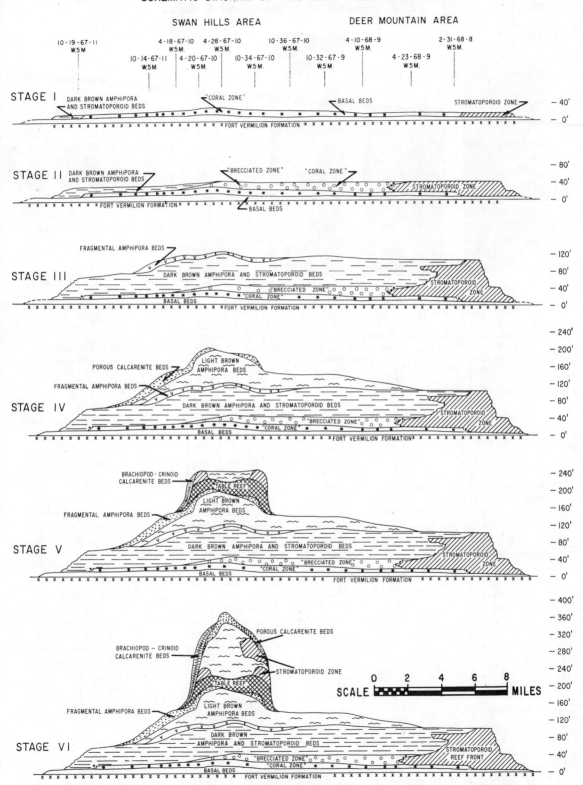

SCHEMATIC DIAGRAM OF THE SWAN HILLS REEF GROWTH

27

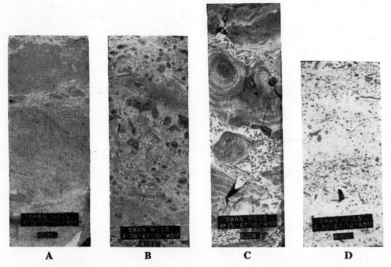

A   B   C   D

FIG. 17.—**A,** Fine-grained, dark-brown to black argillaceous beds of basal Beaverhill Lake, 3 ft below "Coral zone." Fine clasts of similar material can be seen scattered throughout, suggesting reworking of beds. **B,** *Thamnopora*-type corals in matrix of buff calcarenite, representing "Coral zone" of Swan Hills Formation. Unit forms lowermost productive bed in Swan Hills field. In this sample "Coral zone" has 8.2 percent porosity and 3.6 md permeability. **C,** Broken, bulbous and massive stromatoporoids of "Brecciated zone." Fine *Amphipora* in dark-brown micritic matrix fill interstices between larger stromatoporoid fragments. Some fractured stromatoporoids have calcite infilling. **D,** Dark-brown micritic *Amphipora* beds. Banded appearance results from long axis of *Amphipora* being deposited parallel with normal bedding, and different energy levels which prevailed in carrying *Amphipora* fragments into lagoon.

low turbulent water. The upper boundary with the offreef Waterways Formation is unconformable and appears to be an erosion surface.

Brachiopod-crinoid beds.—The brachiopod-crinoid beds are found most commonly on the west slopes of the Light Brown member, and consist of calcarenites of skeletal origin. In order of decreasing importance the organic components are brachiopods, crinoids, stromatoporoids, and *Amphipora.* The stromatoporoids and *Amphipora* gradually decrease in number, and the calcarenite grades into calcisiltite and fine carbonate mudstone farther away from the main buildup. This brachiopod-crinoid zone with fine calcarenite derived from skeletal material is typical of the reef flanks and is considered to have been deposited in open-marine, aerated water of medium energy. The beds generally are devoid of porosity and therefore have no economic significance.

DEPOSITIONAL HISTORY OF SWAN HILLS FORMATION

In the Swan Hills field area the depositional history of the Swan Hills Formation can be described in six stages, each indicating a major change in sedimentation and environment (Fig. 16). Stages I-III cover the time of Dark Brown member deposition, and Stages IV-VI cover the time of Light Brown member deposition.

Stage I.—The Beaverhill Lake sea gradually transgressed westward from the deepest part of the basin. The area between the Peace River arch and West Alberta ridge was flooded, and the widespread shoal conditions that developed favored organic growth. The depositional environment of associated sediments was similar to that of the offreef Waterways Formation, except for the higher percentage of skeletal material and the first appearance of slightly shallower water organisms (Fig. 17A).

After the appearance of the corals in the upper part of the "Basal beds," the area became densely colonized by frame-building organisms that created a solid base and good foothold for additional organic growth (Fig. 17B). Along the northeast rim, in areas of slightly higher wave energy, a stromatoporoid zone began to develop, whereas farther south along the platform edge in the areas of lower energy, the corals and stromatoporoids were replaced by *Amphipora*-rich beds.

Stage II.—A slight but persistent rise in sea level caused continued stromatoporoid reef growth on the northeast rim. This growth created a broad shelf lagoon behind the

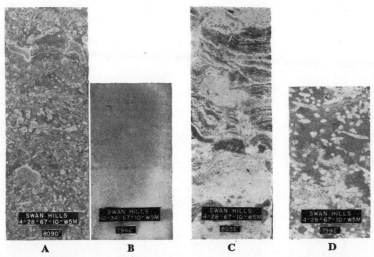

A   B   C   D

FIG. 18.—**A,** Light-brown beds of *Amphipora* bank deposits. In addition to abundant *Amphipora* there are some *Stachyoides* and very few bulbous stromatoporoids in calcarenite matrix. Because of reworking of these beds, *Amphipora* are more randomly oriented and widely scattered. **B,** Fine skeletal reef calcarenite of reworked *Amphipora* bank deposits shows no bedding or recognizable organisms, and is very homogeneous rock. Sample has 9.1 percent porosity and 18 md permeability. **C,** Almost solid framework of massive stromatoporoids bound by algal mats; forms "table reef" (Fig. 5). Calcarenite fills spaces between larger organisms. In sample shown porosity is 12.2 percent with 12 md permeability. **D,** *Amphipora* in light-buff calcarenite. This is typical lagoon deposit just behind reef rim. In sample porosity is 5.5 percent and permeability 10 md.

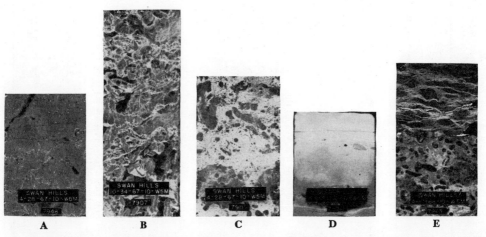

A   B   C   D   E

FIG. 19.—**A,** Pelletoid unfossiliferous carbonate mud deposited in quieter and deeper water of central lagoon. **B,** Stromatoporoid reef-wall material with abundant sparry calcite filling interstices. Organic porosity is 9 percent and permeability is 11 md. **C,** Bioclastic material of "Detrital zone." Reef rubble composed of large and small organic fragments enclosed in dark-brown calcarenite matrix common to top and upper slopes. Porosity is 12.6 percent and permeability 64 md. **D,** Dark-gray, cryptograined, dense, argillaceous, finely bedded calcareous shale of Waterways Formation. **E,** Contact between Swan Hills Formation and overlying Waterways Formation. Close examination of contact shows truncated organisms indicative of erosion before burial. *Boudinage* structure common to some of Waterways beds is present here

Table 1. Parameters of Swan Hills Oil Fields

Fields, Pools	Disc. Date	Name and Loc. Disc. Well	Total Wells Drld. to 12/31/67	Av. Well Depth (Ft)	Prod. Area (Acres)	Maximum Reservoir Thickness (Ft)	Av. Net Pay (Ft)	Av. Por. (%)	Av. Perm. (Md)	Water Sat. (%)	Est. Oil in Place (1,000 Bbl)	Cum. Prod. to 12/31/67 (1,000 Bbl)	Remaining Recoverable Crude Oil (1,000 Bbl)
Swan Hills A and B pools	3/ 2/57	Home et al. Regent 8-11-68-10-W5M	520	8,299	102,479	395	52.8	7.8	20	18.6	1,944,000	83,309	688,621
C pool	3/21/58	Texcan Mic Mac Deer Mtn. 10-14-69-9-W5M	353	7,475	55,058	120	29.8	6.2	5	10.2	553,000	17,684	141,616
Judy Creek A pool	2/25/59	Imperial Judy Creek 16-31-63-10-W5M	184	8,665	28,200	358	66.7	9.3	43	16	809,000	42,018	322,032
B pool	9/ 6/59	Imperial Virginia Hills 10-13-63-12-W5M	75	8,842	12,380	453	58.4	9.0	111	17	256,000	13,123	109,757
Swan Hills South	2/27/59	B.A. Pan Am. Sarah Lake 2-13-65-11-W5M	243	8,345	35,680	405	69.8	8.0	26	18.3	897,600	43,639	348,512
Virginia Hills	1/31/57	Home Union H.B. Virginia Hills 9-20-65-13-W5M	134	9,283	24,350	503	48.4	8.1	35	20.4	450,200	25,288	148,992
Kaybob	4/22/57	Phillips Kaybob 7-22-64-19-W5M	107	9,780	18,000	232	59.9	7.4	23.3	22	300,000	22,346	97,654
Carson Creek North	9/ 6/58	Mobil P.R. Carson N 6-1MU 6-1-62-12-W5M	46										
A pool—oil pool gas cap				8,632 8,580	6,916 3,720		31 11	8.3 8.0	56 15	13	66,000	4,100	24,940
B pool—oil pool gas cap				8,736	12,372 377		51 7.8	9.1 9.9	167 21	21 21	215,000	8,939	74,911
Snipe Lake	10/24/62	S.O.B.C. Snipe Lake 10-21-70-18-W5M	114	8,534	17,297	272	33.3	7.3	35	27	198,000	10,587	66,633
Goose River (B pool excl.)	8/28/63	B.A. Goose River 10-4-67-18-W5M	28	9,185	7,700	200	49	8.2	103	19	145,000	2,613	20,587
Freeman	10/31/62	H.B. Union Home Freeman 2-1 2-1-66-13-W5M	20	9,184	5,390	160	29	5.9	20	25	40,400	874	3,934
Ante Creek	10/15/62	Atlantic Ante Creek 4-7 4-7-65-23-W5M	24	11,270	7,510	260	25.3	6.3	6	22	34,800	1,838	3,730
Judy Creek South	3/19/60	Mobil Carson Creek 14-31 14-31-62-11-W5M	8	8,925	3,009	136	22.6	6.3	24.4	25	15,000	557	2,443
Ethel	1/28/64	Mobil Atlantic Ethel 10-11 10-11-67-8-W5M	4	7,522	1,289	87	23.6	5.7	10.2	17	8,100	19	62

wave-resistant reef front. In the quieter and more restricted waters behind the reef, the less wave-resistant bulbous stromatoporoids of the "Brecciated zone" flourished (Fig. 17C). Farther west, in the quiet, semistagnant waters of the lagoon, *Amphipora* beds were deposited in precipitated carbonate mud.

Stage III.—The rigid stromatoporoid reef wall continued to grow on the northeast side of the carbonate bank, while *Amphipora*-rich beds were deposited in the shelf lagoon behind the reef front (Fig. 17D). The writers assume that carbonate-bank growth ceased when slight eastward tilting deepened the water above the reef-front area, and at the same time part of the carbonate bank became emergent on the west. Brief exposure of the east-central part of the carbonate platform caused reworking of the upper part of the bank in the Swan Hills area and created a thin fragmental zone on this part of the carbonate platform. This marked the termination of deposition for the Dark Brown member and provided a substratum for further organic buildup on the slightly higher, emergent western side.

Stage IV.—After the submergence of the stromatoporoidal reef front, *Amphipora* grew abundantly in the quieter, slightly deeper water of the lagoon bank. The periods of quiescence were interrupted by frequent storms, which transported *Amphipora* fragments into localized carbonate-bank deposits that formed loci for the reef growth of Stage V (Fig. 18A). The presence of porous calcarenite near the top and along the flanks of these bank deposits suggests near emergence at this time, and deposition ceased (Fig. 18B).

Stage V.—Stage V is the "Table reef" which developed on the beds of Stage IV. A temporary stillstand of the sea permitted lateral growth of the organic lattice, which resulted in this widespread, fairly homogeneous buildup (Fig. 18C). Slow subsidence resulted in the formation of a circular stromatoporoidal reef atoll enclosing a central lagoon (Fig. 18D). Fluctuating water level or occasional storms caused the deposition of small amounts of fragmental material in zones within the lagoon and the reef rubble of the outer slopes. Local emergence probably terminated this stage of growth.

Stage VI.—The final stage of Swan Hills reef development was marked by a return to supratidal conditions. This caused renewed growth of *Amphipora* and the deposition of pelletoid and unfossiliferous carbonate-mud beds (Fig. 19A). Carbonate-mud accumula-

Table 2. Parameters of Swan Hills Gas Fields

Fields, Pools	Disc. Date	Name and Loc., Disc. Well	Total Wells Drilled to 12/31/67	Av. Well Depth (Ft)	Prod. Area (Acres)	Max. Reservoir Thickness (Ft)	Av. Net Pay (Ft)	Av. Por. (%)	Av. Perm. (Md)	Water Sat. (%)	Initial GIP (Bcf)	Marketable Gas Prod. (Bcf)	Remaining Marketable Gas to 12/31/67
Kaybob South	9/11/61	H.B. Union Kaybob 11-27 11-27-62-20-W5M	8	10,560	16,440	325	43	10	294.2	15	670	0	370
Carson Creek	2/26/57	Mobil Oil Whitecourt 12-13 12-13-61-12-W5M											
A pool			8	8,550	15,840	89	20	8	21	20	210	8	142
B pool			5	8,610	6,980	83	24	8	75	20	110	-15	95

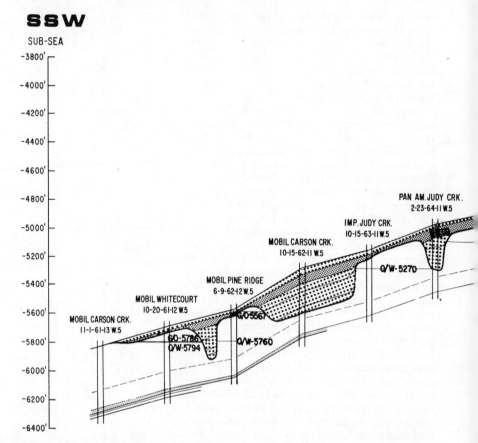

Fig. 20.—SSW-NNE structural cross section through eastern fields (Carson Creek to Swan Hills) shows lateral progression of gas-oil and oil-water interfaces. This is prime example illustrating Gussow's (1954) hypothesis of differential entrapment of hydrocarbons. Vertical scale in feet; sea-level datum. Trace is same as C-C′ (Fig. 12), shown on Figure 21.

tion alternated with the deposition of *Amphipora*-rich beds and a few thin terrigenous mud stringers. Incipient reef growth is present locally (Fig. 19B). Upward organic growth was prevented by emergence and strong erosion. The result of this activity was the formation of calcarenite beds and coarse reef rubble on the top and upper slopes of the buildup. These carbonate clastic rocks are the youngest strata of the Swan Hills Formation (Fig. 19C). Penecontemporaneous deposition of carbonate mud and clay (Waterways Formation) in the offreef areas is indicated (Fig. 19D). Sedimentation proceeded at a slightly slower rate relative to the development of the Swan Hills section.

On the east side of the buildup the contact with the stratified Waterways equivalent of the reef is sharp and unconformable (Fig. 19E). On the west side the brachiopod-crinoid beds,

which show a gradual decrease in grain size and organic content away from the reef, form a transitional zone between the two formations.

At the end of Stage VI, a sudden increase in water depth drowned the reef complex. Waterways-Woodbend clay and carbonate-mud deposits covered most of the area.

Diagenesis of Swan Hills Carbonates

Dolomitization.—The absence of significant dolomitization in the Swan Hills Formation makes it possible to conduct detailed facies, textural, and environmental studies of the reefs. Most of the formation is in the initial stage of compaction-current dolomitization, which involves decreased porosity and reduced pore volume. Where dolomitization has occurred, it generally plugs void space of primary organic and matrix porosity. Presumably the magne-

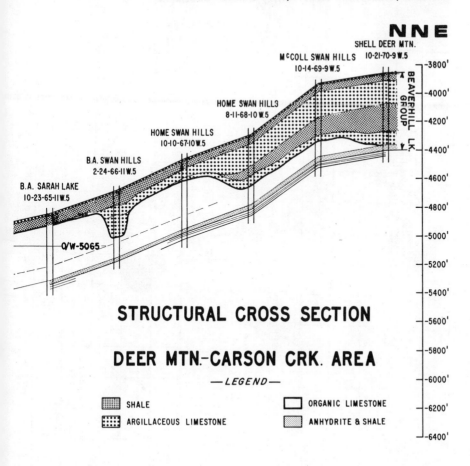

NNE

SHELL DEER MTN.

M^cCOLL SWAN HILLS   10-21-70-9 W.5

10-14-69-9 W.5

HOME SWAN HILLS
8-11-68-10 W.5

HOME SWAN HILLS
10-10-67-10 W.5

B.A. SWAN HILLS
2-24-66-11 W.5

B.A. SARAH LAKE
10-23-65-11 W.5

O/W-5065

STRUCTURAL CROSS SECTION

DEER MTN.-CARSON CRK. AREA

—LEGEND—

SHALE   ORGANIC LIMESTONE

ARGILLACEOUS LIMESTONE   ANHYDRITE & SHALE

sium ions were brought to the site of precipitation by percolating waters after lithification.

Another form of dolomitization is associated with the organisms; *e.g.*, the axial canals in *Amphipora* commonly contain dolomite infilling in varied amounts. Generally this dolomite infilling is associated with calcite crystals. Presumably the source of magnesium was the organisms, or magnesium ions from areas of higher ion concentration which migrated to and accumulated in the fossils.

A less common form of dolomite is a scattering of perfect dolomite rhombs, about 20–50 μ in diameter, occupying space once filled by much finer calcite grains and microorganisms in a generally dense matrix.

Complete dolomitization of the Swan Hills Formation is observed south and west of the Kaybob area, and south of the Carson Creek gas field, where it is associated mainly with the upper part of the Light Brown member. The reason for this complete dolomitization is not known to the writers, but is assumed to be related to tectonism.

Silicification.—Scarce light-gray to brown-gray dolomitic chert lenses are found in the Swan Hills Formation. Thickness ranges from 1 to 3 in. The material partly replaces skeletal and nonskeletal grains, and possibly was precipitated from chemical solutions of organic origin.

Recrystallization.—Recrystallization is here defined as the formation of new mineral grains in the limestone in the solid state, without the introduction of other elements. Recrystallization is not very important in the Swan Hills area, but it has been observed in a few places in the upper part of the Light Brown member,

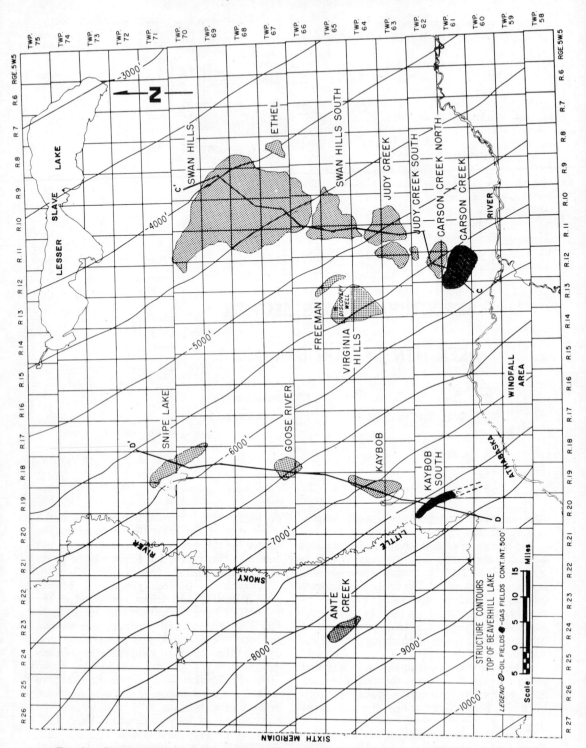

Fig. 21.—Structural contours, datum top of Beaverhill Lake, show southwesterly regional dip. Lack of differential compaction is shown by absence of contour deviation as contours pass through areas of reef buildup. CI = 500 ft. Depth subsea. Shows locations of Figures 12, 13, and 20.

generally in association with the finer grained sediments of the lagoon facies. The matrix is converted into a limestone of saccharoid texture, and the original structure of the *Amphipora* and other organisms is blurred or lost.

Fracturing.—A study was made of numerous cores from the Swan Hills field and the fractured intervals were recorded. Generally, fractures are very scarce in the Dark Brown member, but are more common in the upper part of the Light Brown member.

The matrix of the Dark Brown member consists of very fine-grained calcarenite and calcilutite, which commonly are recemented by secondary calcite. Recementation caused considerable compaction. The lithologic character of the Light Brown member reflects sedimentation in high-energy waters; the matrix ranges in lithologic character from loosely consolidated, fine-grained calcarenite to calcirudite which yielded to stress more easily by fracturing.

Detailed cross-section studies of the Beaverhill Lake Group show that this limestone interval thins slightly over areas of maximum reef buildup. The reduction in thickness probably resulted from the collapse of porous sections accompanied by the development of fractures. Most fractures are vertical or high-angle oblique and many of the fracture planes are lined with secondary calcite crystals. The importance of these fractures is that they may permit communication between otherwise isolated porous zones within the Light Brown member.

RESERVOIR CHARACTERISTICS

Reservoir parameters for each field are shown in Tables 1 and 2. Within the map area, approximately 377,000 acres is underlain by productive Swan Hills Formation. In individual wells net pay thickness is as much as 250 ft. However, the best field average is 69.8 ft and the poorest 22.6 ft. Field-weighted average-po-

Table 3. Properties of Natural Gas, Carson Creek and Kaybob South Fields

Field	Sp. Gr. Gas	Recoverable NGL Content (Bbl/MMcf)[1]	Recoverable Sulfur Content (Long Tons/MMcf Raw Gas)
Carson Ck. A	0.930	131.48	—
B	0.972	142.17	—
Kaybob South	1.0096	107	6.2

[1] Based on raw gas volume with 85 percent recovery of propane, 95 percent recovery of butane, and 100 percent recovery of heavier ends.

Table 4. Beaverhill Lake Fields, Oil Gravity and Saturation Pressures

Field	API Gravity Produced Liquid (°API)	BHT (°F)	Ps @ BHT (psig)	Ps @ 225° F (psig)
Ante Creek	46	233	4,140	4,116
Carson Creek oil leg[1]	43	198	3,767	3,848
Carson Creek gas	61	198	(3,850 DP)[2]	—
Carson Creek N. A	44	187	3.391	3,493
Carson Creek N. B[3]	44	191	3,312	3,414
Freeman	40	221	1,738	1,750
Goose River	40	233	2,052	2,028
Judy Creek A	43	205	2,290	2,350
Judy Creek B	43	206	2,940	2,997
Kaybob S.	60	238	(3,643 DP)	—
Kaybob	43	235	3,019	2,989
Snipe Lake	37	183	1,300	1,426
Swan Hills S.	42	225	2,219	2,219
Swan Hills (Main)	43	225	1,820	1,820
Swan Hills (Inverness)	42	211	1,726	1,768
Swan Hills C	42	189	1,370	1,478
Virginia Hills	38	218	1,812	1,833

[1] There is a thin (8 ft) noncommercial oil leg in this reservoir.
[2] DP denotes dew-point pressure for gas fields.
[3] Reservoir has a small gas cap but oil leg appears undersaturated.

rosity values range from a low of 5.7 percent to a high of 10 percent, whereas weighted average horizontal permeability values for the fields range from 5 to 167 md.

All of the oil pools are undersaturated and there is a general increase in API gravity toward the gas-condensate pools. An odd situation is present in the Carson Creek North field, where both the A and B pools have a gas cap but samples of the reservoir oil indicate undersaturation. Snipe Lake has the lowest gravity of 37° and Ante Creek has the highest of 46°.

Water saturation ranges from 10 to 30 percent, but generally is about 20 percent. The structural relations and the differing water tables along the eastern reef trend are depicted in Figure 20.

Oil is carried to market by the Peace River pipeline and the Federated pipeline. Posted wellhead price is approximately $2.55/bbl (Canadian).

In several fields, the presence of separate pools has been established. Four factors account for this situation:

1. Presence of oil in separate reef facies within the same reef complex (the best example is Swan Hills).
2. Vertical separation by an impermeable green shale barrier, such as is found in the Carson Creek and Carson Creek North fields.
3. Horizontal and vertical permeability barriers caused by completely "tight" reef facies; *e.g.*, the Judy Creek–Judy Creek South relation.

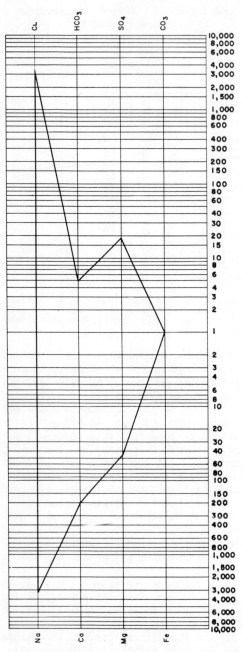

FIG. 22.—Logarithmic plot of typical water analysis, Swan Hills Formation, in milligram equivalents per unit. There is considerable uniformity in chemical composition of formation waters from Swan Hills fields throughout area. This illustration is thought to represent typical logarithmic pattern of a Swan Hills Formation water analysis.

4. Scour or surge channels between reef masses typical of the eastern reef chain, as shown on Figure 14. Dissection of this chain is apparent from the number of separate pools along the trend.

The Light Brown member of the Swan Hills Formation has the best reservoir characteristics. In fact, it is only in the Swan Hills field that the Dark Brown member and the "Coral zone" are of economic importance. There the porous beds of the Dark Brown and Light Brown members are grouped and designated as one pool for proration purposes. The "Coral zone" is separated by the impermeable carbonate of the Dark Brown member and forms the B pool of the field. Probably the principal reason for production in the Dark Brown member is the lack of definite water table, which suggests that the entire basal part of the Swan Hills Formation is above the oil-water interface as calculated from regional pressure data.

Depth of drilling is controlled primarily by the position of the well in the Alberta basin. Regional dip is southwest at approximately 40 ft/mi in the eastern part of the area and steepens to 50 ft/mi in the western part (Fig. 21). Accordingly, drilling depth ranges from 7,475 ft at the northeast end of the Swan Hills field to more than 11,000 ft in the Ante Creek area.

Reservoir Fluids

Gas.—The primary difference in gas composition between the two Beaverhill Lake gas fields (Table 2; Carson Creek and Kaybob South) is in the sulfur content. Gas from Carson Creek is sweet, whereas Kaybob South gas has 16.59 percent H_2S by volume. Both fields are rich in natural gas liquids. Table 3 gives the most important gas properties. Small gas caps are present in the Carson Creek North A and B pools.

Oil.—Oil properties differ somewhat among fields. This is to be expected because of the many reservoirs, each of which has different physical properties. Generally, the Beaverhill Lake reservoirs contain an undersaturated paraffin-base crude, with a low sulfur content ranging from a trace to 0.42 percent. The variations in API gravity are given in Table 4, together with the bottomhole temperature and saturation pressure for each field. The saturation pressures also have been adjusted to a common temperature of 225° F, to permit easier correlation of the reservoir-fluid properties.

Water.—There is an oil-water or gas-water

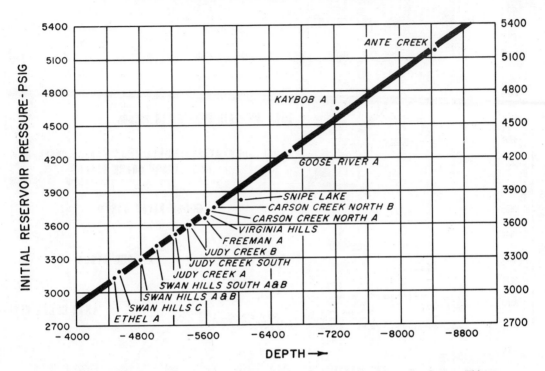

FIG. 23.—Plot of initial pressure *versus* datum (depth subsea), Beaverhill Lake Group. This pressure-depth plot of Swan Hills fields shows close relation of initial pressures, suggesting common pressure system for area. Data from Oil and Gas Conservation Board, Calgary (1967).

interface in most of the developed fields. Notable exceptions are Swan Hills, Ethel, Kaybob, and Freeman. Where a variable water table is reported, it generally can be attributed to a change in lithologic character within the interval in which the water table normally would be. A very few water occurrences which are difficult to explain have been found in the Swan Hills field. Probably rock-geometry is involved (Stout, 1964).

The chemical compositions of different formation waters can be compared by plotting the components as a logarithmic pattern. The figures used are milligram equivalents per unit, obtained by multiplying parts per million (mg/l) by the following factors: Na + K × 0.0435; Ca × 0.0499; Mg × 0.0822; SO$_4$ × 0.0208; Cl × 0.0282; CO$_3$ × 0.0333; and HCO$_3$ × 0.0164.

Water analyses from several Beaverhill Lake fields were plotted in this manner and a typical logarithmic pattern for Swan Hills Formation waters is shown in Figure 22.

PRESSURE RELATIONS

A pressure-depth plot has been made by the Oil and Gas Conservation Board using the original pressures for several of the Swan Hills fields. The close relation of these initial pressures suggests a common pressure system (Fig. 23).

A similar detailed plot was made by selecting the most reliable pressure data from the individual wells in the gas, oil, and water phases of the Swan Hills Formation. To it have been added pressure data from the fluid phases of the widespread Gilwood Sandstone, which is the highest reservoir rock below the Swan Hills Formation. The data indicate a connection between these two reservoirs (Fig. 24).

It is not the writers' intention to speculate on the various stages of fluid migration or to explain the reason for the Gilwood-Swan Hills relation. Many factors are involved, such as rate of compaction, thickness of overburden, regional tilting, fracturing, juxtaposition of porous beds, *etc.* In some areas the Beaverhill

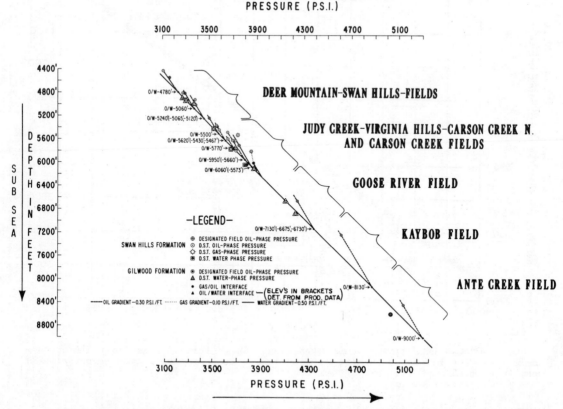

FIG. 24.—Pressure-depth plot, Beaverhill Lake–Gilwood Sandstone system, illustrates relation of Swan Hills oil and gas pools to a common reservoir system. Pressures measured within water phase in both Gilwood Sandstone and Swan Hill Formation reservoirs also relate to single system, suggesting that Swan Hills reservoir continuity is achieved through widespread Gilwood aquifer

Lake directly overlies the Gilwood Sandstone, or the "granite wash" with which the Gilwood Sandstone merges in the vicinity of the Peace River arch. Presumably these areas constitute the principle points for pressure communication.

On Figure 24 two oil-water interfaces are shown for some of the fields. One is the hypothetical interface based on the fluid-phase pressure data relation and represents the level at which the pore space is filled with 100 percent formation water. The other (in brackets) is the operating field oil-water interface, which has been established from production, drill-stem tests, and log interpretations. Field oil-water interfaces generally are established in the transition zone at a point where water-free production is obtained. This probably accounts for the differences between the calculated and reported oil-water contacts.

Logging

The most common logging devices used in Swan Hills fields are the induction electric log, microcaliper log, and gamma-sonic log. The microlog is an excellent tool for establishing effective net pay, whereas the sonic log has been used widely for calculating porosity in uncored wells and sections of uncored wells.

Secondary Recovery

Without exception, early production history of the field and the pools producing oil from the Swan Hills Formation presaged rapid pressure decline. None of the fields has been characterized by a strongly active water drive, and all indicated a low recovery of approximately 16 percent of the oil in place based on primary depletion. The primary-depletion recovery mechanism is rock and fluid expansion down to

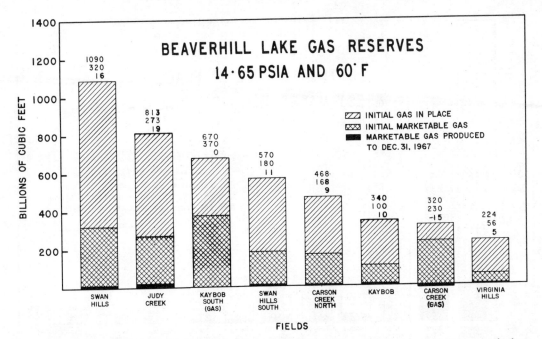

Fig. 25.—Bar diagram of Beaverhill Lake gas reserves shows initial gas in place, marketable gas, and where significant gas is produced for each of principal Swan Hills fields. Gas produced from Carson Creek North field was reinjected into Carson Creek gas field and accounts for negative value.

the bubble point, followed by a relatively inefficient solution gas drive.

Concern of the operating companies and the Alberta Oil and Gas Conservation Board resulted in exhaustive reservoir studies to find the most efficient means of maintaining field pressures above the bubble point, and thus obtain the maximum ultimate recovery of oil. The findings prompted unitization of most of the fields and, with one exception, water injection has been selected as the method of secondary recovery. For the Ante Creek field, a miscible flood is planned because of the higher initial reservoir pressure and solution gas-oil ratio resulting from greater depth of burial.

To the writers' knowledge, for every principal Swan Hills oil accumulation a secondary-recovery scheme is planned or is in effect. Secondary-recovery techniques are expected to result in recoveries of 35–60 percent of the original oil in place.

Water is obtained from surface sources, with approval of the Water Resources Branch of the Department of Agriculture, and is treated chemically for purity and bacteria control before injection. In most cases a line-drive method is adopted using downdip wells and injecting water below the field oil-water interface.

Where no field water table is present and it is necessary to inject into previously oil-producing wells, injection characteristics generally are not as good.

RESERVES

Gas.—Dissolved gas provides most gas reserves from the Swan Hills Formation. This, plus the reserves at Carson Creek and Kaybob South gas fields, accounted for a total of 4.512 trillion ft[3] initial raw gas in place to December 31, 1967. Conservation practices that require gathering most of the casinghead gas for marketing or reinjection into the formation result in an estimated initial marketable gas reserve of 1.097 trillion ft[3] from this source. Addition of the estimated marketable reserves from the two gas fields places the total for the area at 1.697 trillion ft[3].

Figure 25 shows the initial gas in place, marketable gas, and gas produced for each of the Swan Hills fields. Gas produced from the Carson Creek North field has been reinjected into the Carson Creek gas field and accounts for the negative value shown for marketable gas produced for that field.

Oil.—Total recoverable oil for all the fields producing from the Swan Hills Formation has

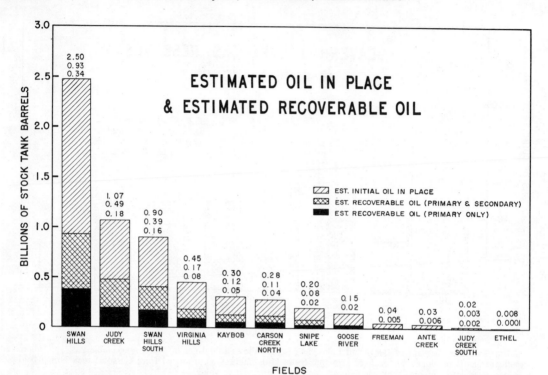

FIG. 26.—Estimated oil in place attributable to each of Swan Hills fields. Estimated recoverable oil by primary recovery and combination of primary- and secondary-recovery methods also is shown.

been estimated at 2.331 billion bbl to December 31, 1967. If only light- and medium-gravity crudes are considered, this figure amounts to 24.78 percent of the total known recoverable oil reserves for the Province of Alberta. The estimated oil in place attributable to each field, together with reserves expected to be produced by primary recovery and a combination of primary- and secondary-recovery methods, is shown in Figure 26.

The history of the petroleum industry in the Province of Alberta has been cyclical, characterized by a series of intermittent major discoveries. The effect of additional reserves from a major strike, such as that of the Swan Hills Formation, is shown in Figure 27.

ECONOMIC IMPACT

An estimate of the total funds spent within the map area after the initial discovery would require compiling figures on seismic work, road building, drilling and completion of wells, townsites, land acquisition, and pipelines. Though such a study is beyond the scope of this paper, it is interesting to examine the factor of land acquisition.

In Alberta, exclusive of early settlement areas, the Crown owns the petroleum and natural gas rights. These rights are disposed of by a system of closed bidding at periodic government sales. There were no freehold lands within the Swan Hills area. In the 2 years before the discovery, approximately $140,291 had been spent for the acquisition of petroleum and natural gas rights within the map area. In the years 1957–1967 this figure rose to $184,034,918.

SELECTED REFERENCES

Andrichuk, J. M., 1958, Stratigraphy and facies analysis of Upper Devonian reefs in Leduc, Stettler, and Redwater areas, Alberta: Am. Assoc. Petroleum Geologists Bull., v. 42, no. 1, p. 1–93.
Bassett, H. G., 1961, Devonian stratigraphy, central Mackenzie River region, Northwest Territories, Canada, in Geology of the Arctic, v. 1: Toronto, Toronto Univ. Press, p. 481–498.
——— and J. G. Stout, 1967, The Devonian stratigraphy of western Canada, in International Symposium on the Devonian System, v. 1: Alberta Soc. Petroleum Geologists, p. 717–752.
Beales, F. W., 1957, Bahamites and their significance in oil exploration: Alberta Soc. Petroleum Geologists Jour., v. 5, no. 10, p. 227–231.
——— 1958, Ancient sediments of Bahaman type:

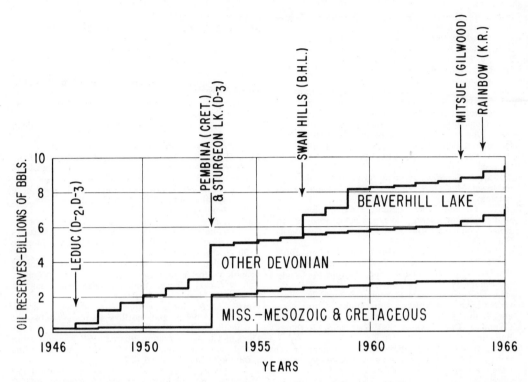

FIG. 27.—Alberta's cumulative ultimate proved oil reserves distributed according to year of discovery to Dec. 13, 1966. Share of Alberta's cumulative ultimate proved oil reserves attributable to Swan Hills fields is shown. By end of 1966 Swan Hills fields accounted for 25.87 percent of Alberta total.

Am. Assoc. Petroleum Geologists Bull., v. 42, no. 8, p. 1845–1880.

———— 1960, Limestone peels: Alberta Soc. Petroleum Geologists Jour., v. 8, no. 4, p. 132–135.

Beard, D. E., 1959, Selective solution in the Devonian Swan Hills Member: Alberta Soc. Petroleum Geologists Jour., v. 7, no. 7, p. 163–164.

Belyea, H. R., 1952, Notes on the Devonian System of the north central plains of Alberta: Canada Geol. Survey Paper 52–27, 66 p.

———— 1955, Correlations in the Devonian of southern Alberta: Alberta Soc. Petroleum Geologists Jour., v. 3, no. 9, p. 151–156.

———— 1964, Upper Devonian, pt. II, in Geological history of Western Canada: Alberta Soc. Petroleum Geologists, p. 66–81.

———— and A. W. Norris, 1962, Middle Devonian and older Paleozoic formations of southern District of Mackenzie and adjacent areas: Canada Geol. Survey Paper 62–15, 82 p.

Bonham-Carter, C. F., 1963, A study of microscopic components of the Swan Hills Devonian reef: Unpub. M.S. thesis, Toronto Univ.

Brown, P. R., 1963, Some algae from the Swan Hills reef: Bull. Canadian Petroleum Geology, v. 11, no. 2, p. 178–182.

Burwash, R. A., H. Baadsgaard, Z. E. Peterman, and G. H. Hunt, 1964, Precambrian, chap. 2, in Geological history of Western Canada: Alberta Soc. Petroleum Geologists, p. 14–19.

Cameron, A. E., 1918, Explorations in the vicinity of Great Slave Lake: Canada Geol. Survey Summ. Rept., pt. C, 1917, p. 21–28.

———— 1922, Hay and Buffalo Rivers, Great Slave Lakes, and adjacent country: Canada Geol. Survey Summ. Rept., pt. B, 1921, p. 1–44.

Campbell, N., 1950, The Middle Devonian in the Pine Point area, N.W.T.: Geol. Assoc. Canada Proc., v. 33, p. 87–96.

Carozzi, A. V., 1961, Reef petrography in the Beaverhill Lake Formation, Upper Devonian, Swan Hills area, Alberta, Canada: Jour. Sed. Petrology, v. 31, no. 4, p. 497–513.

Century, J. R. (ed.), 1966, Oil fields of Alberta, supplement: Alberta Soc. Petroleum Geologists, 136 p.

Clark, D. L., and R. L. Ethington, 1965, Conodont biostratigraphy of part of the Devonian of the Alberta Rocky Mountains: Bull. Canadian Petroleum Geology, v. 13, no. 3, p. 382–389.

Crickmay, C. H., 1954, Paleontological correlation of Elk Point and equivalents, in Western Canada sedimentary basin: Am. Assoc. Petroleum Geologists, p. 143–158.

———— 1957, Elucidation of some Western Canada Devonian formations: Imperial Oil Ltd., unpub. rept.

De Mille, G., 1958, Pre-Mississippian history of the Peace River arch: Alberta Soc. Petroleum Geologists Jour., v. 6, no. 3, p. 61–68.

Edie, E. W., 1961, Devonian limestone reef reservoir, Swan Hills oil field, Alberta: Canadian Inst. Mining and Metallurgy Trans., v. 64, p. 278–285.

Fischbuch, N. R., 1960, Stromatoporoids of the Kaybob reef, Alberta: Alberta Soc. Petroleum Geologists Jour., v. 8, p. 113–131.

—— 1962, Stromatoporoid zones of the Kaybob reef, Alberta: Alberta Soc. Petroleum Geologists Jour., v. 10, no. 1, p. 62–72.

Folk, R. L., 1959, Practical petrographic classification of limestones: Am. Assoc. Petroleum Geologists Bull., v. 43, no. 1, p. 1–38.

Fong, G., 1959, Type section Swan Hills Member of the Beaverhill Lake Formation: Alberta Soc. Petroleum Geologists Jour., v. 7, no. 5, p. 95–108.

—— 1960, Geology of Devonian Beaverhill Lake Formation, Swan Hills area, Alberta: Am. Assoc. Petroleum Geologists Bull., v. 44, no. 2, p. 195–209.

Galloway, J. J., 1960, Devonian stromatoporoids from the lower Mackenzie Valley of Canada: Jour. Paleontology, v. 34, no. 4, p. 620–636.

Gray, F. F., and J. R. Kassube, 1963, Geology and stratigraphy of Clarke Lake gas field, northeastern British Columbia: Am. Assoc. Petroleum Geologists Bull., v. 47, no. 3, p. 467–483.

Grayston, L. D., D. F. Sherwin, and J. F. Allan, 1964, Middle Devonian, chap. 5, in Geological history of Western Canada: Alberta Soc. Petroleum Geologists, p. 49–59.

Griffin, D. L., 1965, The Devonian Slave Point, Beaverhill Lake, and Muskwa Formations of northeastern British Columbia and adjacent areas: British Columbia Dept. Mines and Petroleum Resoures Bull., no. 50, 90 p.

Gussow, W. C., 1954, Differential entrapment of oil and gas, a fundamental principle: Am. Assoc. Petroleum Geologists Bull., v. 38, no. 5, p. 816–853.

Guthrie, D. C., 1956, Gilwood Sandstone in the Giroux Lake area, Alberta: Alberta Soc. Petroleum Geologists Jour., v. 4, no. 10, p. 227–231.

Hriskevich, M. E., 1966, Stratigraphy of Middle Devonian and older rocks of Banff Aquitaine Rainbow West 7–32 discovery well, Alberta: Bull. Canadian Petroleum Geology, v. 14, no. 2, p. 241–265.

Illing, L. V., 1959, Deposition and diagenesis of some upper Palaeozoic carbonate sediments in Western Canada: 5th World Petroleum Cong., Sec. 1, Paper 2, p. 23–52.

Imperial Oil Limited, Geological Staff, 1950, Devonian nomenclature in the Edmonton area, Alberta, Canada: Am. Assoc. Petroleum Geologists Bull., v. 34, no. 9, p. 1807–1825.

Jenik, A. J., 1965, Facies and geometry of the Swan Hills Member, Alberta: Unpub. M.Sc. thesis, Alberta Univ.

Koch, N. G., 1959, Correlation of the Devonian Swan Hills Member, Alberta: Unpub. M.Sc. thesis, Alberta Univ.

Kramers, J. W., and J. E. Lerbekmo, 1967, Petrology and mineralogy of Watt Mountain Formation, Mitsue-Nipisi area, Alberta: Bull. Canadian Petroleum Geology, v. 15, no. 3, p. 346–378.

Law, J., 1955, Rock units of northwestern Alberta: Alberta Soc. Petroleum Geologists Jour., v. 3, no. 6, p. 81–83.

Leavitt, E. M., 1966, The petrology, paleontology and geochemistry of the Carson Creek North reef complex, Alberta: Unpub. Ph.D. thesis, Alberta Univ.

LeBlanc, R. J., and J. G. Breeding (eds.), 1957, Regional aspects of carbonate deposition: Soc. Econ. Paleontologists and Mineralogists Spec. Pub. 5, 178 p.

Loranger, D. M., 1965, Devonian paleoecology of northeastern Alberta: Jour. Sed. Petrology, v. 35,

no. 4, p. 818–838.

McGehee, J. R., 1949, Pre-Waterways Paleozoic stratigraphy of Alberta plains: Am. Assoc. Petroleum Geologists Bull., v. 33, no. 4, p. 603–613.

McGill, P., 1966, Ostracods of probable late Givetian age from Slave Point Formation, Alberta: Bull. Canadian Petroleum Geology, v. 14, no. 1, p. 104–133.

McGrossan, R. G., and R. P. Glaister (eds.), 1964, Geological history of Western Canada: Alberta Soc. Petroleum Geologists, 232 p.

McLaren, D. J., and E. W. Mountjoy 1962, Alexo equivalents in the Jasper region, Alberta, Canada: Canada Geol. Survey Paper 62–63, 36 p.

Mound, M. C., 1966, Late Devonian conodonts from Alberta subsurface (abs.): Am. Assoc. Petroleum Geologists Bull., v. 50, no. 3, p. 628.

Moyer, G. L., 1964, Upper Devonian, pt. I, chap. 6, in Geological history of Western Canada: Alberta Soc. Petroleum Geologists, p. 60–66.

Murray, J. W., 1964, Some stratigraphic and paleoenvironmental aspects of the Swan Hills and Waterways Formation, Judy Creek, Alberta, Canada: Unpub. Ph.D. thesis, Princeton Univ.

—— 1966, An oil producing reef-fringed carbonate bank in the Upper Devonian Swan Hills Member, Judy Creek, Alberta: Bull. Canadian Petroleum Geology, v. 14, no. 1, p. 1–103.

Norris, A. W., 1963, Devonian stratigraphy of northeastern Alberta and northwestern Saskatchewan: Canada Geol. Survey Mem. 313, 168 p.

Oil and Gas Conservation Board, 1967, Pressure-depth and temperature-depth relationships, Alberta crude oil pools: OGCB Rep. 67–22.

—— 1968, Reserves of crude oil, gas, natural gas liquids, and sulphur, Province of Alberta: OGCB Rept. 68–18; 175 p.

Porter, J. W., and J. G. C. M. Fuller, 1964, Ordovician-Silurian, pt. 1, chap. 4, in Geological History of Western Canada: Alberta Soc. Petroleum Geologists, p. 34–42.

Russel, L. S., 1967, Palaeontology of the Swan Hills area, north-central Alberta: Toronto Univ. Press, Life Sci. Contr. no. 71, 31 p.

Stout, J. L., 1964, Pore geometry as related to carbonate stratigraphic traps: Am. Assoc. Petroleum Geologists Bull., v. 48, no. 3, p. 329–337.

Thomas, G. E., and H. S. Rhodes, 1961, Devonian limestone bank-atoll reservoirs of the Swan Hills area, Alberta: Alberta Soc. Petroleum Geologists Jour., v. 9, no. 2, p. 29–38.

Uyeno, T. T., 1967, Conodont zonation, Waterways Formation (Upper Devonian), northeastern and central Alberta: Canada Geol. Survey Paper 67–30, 20 p.

Van Hees, H., 1958, The Meadow Lake escarpment—its regional significance to lower Paleozoic stratigraphy, in 1st Internat. Williston Basin Symposium: North Dakota Geol. Soc. and Saskatchewan Geol. Soc., p. 131–139.

—— and F. K. North, 1964, Cambrian, chap. 3, in Geological History of Western Canada: Alberta Soc. Petroleum Geologists, p. 20–33.

Walker, C. T., 1957, Correlations of Middle Devonian rocks in western Saskatchewan: Saskatchewan Dept. Mineral Resources Rept. 25, 59 p.

Warren, P. S., 1933, The age of the Devonian limestone at McMurray, Alberta: Canadian Field-Naturalist, v. 47, no. 8, p. 148–149.

—— 1957, The Slave Point Formation: Edmonton Geol. Soc. Quart., v. 1, no. 1, p. 1–2.

White, R. J. (ed.), 1960, Oil fields of Alberta: Alberta Soc. Petroleum Geologists, 272 p.

BULLETIN OF THE AMERICAN ASSOCIATION OF PETROLEUM GEOLOGISTS
VOL. 44, NO. 5 (MAY, 1960) PP. 569-588, 9 FIGS., 5 PLATES

FACIES AND POROSITY RELATIONSHIPS IN SOME MISSISSIPPIAN CARBONATE CYCLES OF WESTERN CANADA BASIN[1]

G. E. THOMAS[2] AND R. P. GLAISTER[2]
Calgary, Alberta, Canada

ABSTRACT

Case histories of textural and reservoir analyses of two Mississippian carbonate cycles of the Western Canada basin are presented to illustrate the relationships of grain, matrix, and cement variants of carbonate rocks to porosity and permeability determinations.

Large stratigraphic oil pools have been discovered, at or near the Paleozoic subcrop of the Mississippian "Midale" carbonate cycle, in southeastern Saskatchewan. Apart from scattered, vuggy, algal-encrusted strand line deposits, most of the carbonates of the "Midale" producing zone consist of skeletal and oölitic limestones which have a finely comminuted, commonly dolomitized, limestone matrix with intergranular and chalky porosity. Effective reservoir porosity is controlled by the relative distribution and grain size of this matrix.

Major hydrocarbon (oil and gas) reserves have been found in the Mississippian "Elkton" carbonate cycle, both in the Foothills Belt and along the subcrop, in southwestern Alberta. Effective reservoir material of this cycle was found to consist mainly of the dolomitized equivalent of an originally coarse skeletal limestone, with a variable amount of generally porous, finely comminuted (granular) skeletal matrix. Primary porosity was very important in the control of dolomitization which probably began with the replacement of this matrix by euhedral rhombohedrons and finally affected the coarse skeletal material (now generally indicated by leached fossil cast outlines). These porous dolomites grade laterally in a predictable way into tight, relatively non-dolomitized, well sorted, coarse skeletal limestones with original high interfragmental porosity now completely infilled with clear crystalline calcite. This lithification by cementation took place early in the history of carbonate sedimentation of this area and before secondary dolomitization processes took effect.

INTRODUCTION

During the past 10 years there has been an increasing demand from industry, national geological organizations, and universities to organize the classification of carbonate rocks into a single, moderately detailed system of nomenclature which will be understood and used by all concerned. Unfortunately, it now appears that this demand will soon be met by not one but by a plethora of carbonate rock breakdowns.

Any proposed carbonate rock classification which does not attempt to give an explanation for the variances in reservoir void space in limestone or dolomite sequences will not satisfy the requirements of an oil geologist or reservoir engineer. Effective porosity isopach and allied carbonate textural maps will be essential to future exploration and exploitation programs in the Western Canada basin, because most hydrocarbons discovered to date are contained in carbonate stratigraphic traps of organic reef or clastic origin.

[1] Manuscript received, August 6, 1959.

[2] Geologists with Imperial Oil Limited. Thanks are due to Imperial Oil Limited for permission to publish this paper.

The valuable assistance rendered by J. W. Young of Imperial's Regina Division in the "Midale" carbonate study is gratefully acknowledged.

It is fortunate that carbonate sedimentation is so sensitive to environmental conditions that widely various textures and structures result, each of which records the stamp of the depositional conditions which produced it. The full significance of carbonate rock textures in environmental interpretation and in evaluation of potential reservoir zones is becoming more and more apparent.

In this report, case histories of textural and reservoir analyses of two Mississippian carbonate cycles of the Western Canada basin are presented to illustrate the relation between the occurrence and petrographic nature of what constitutes an effective carbonate reservoir rock and the framework of carbonate sedimentation. Various types of carbonate rock pores, which are known to be characteristic of many limestone and dolomite sections, are described and evaluated with respect to effective porosity. These associations of textural type and porosity development are not found exclusively in one area. The descriptions here should be helpful in the recognition and mapping of such associations in other places. Figure 1 is an index map of the areas of study in Saskatchewan and Alberta, Canada. A generalized Mississippian correlation chart of these two areas is shown in Figure 2.

INDEX MAP
TO
AREAS OF STUDY
ALBERTA - SASKACHEWAN CANADA
100 MILES
EDMONTON
EDGE OF DISTURBED BELT
EROSIONAL LIMITS OF MISSISSIPPIAN STRATA
CALGARY
REGINA
B. C.
ALTA.
SASK.
MONTANA
GLASGOW
WILLISTON
N. DAK.
GREAT FALLS

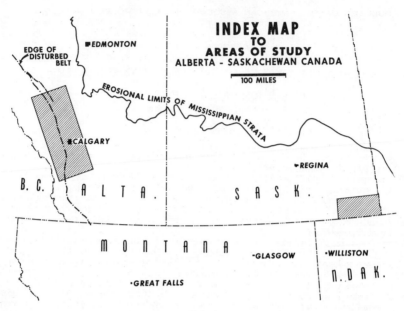

Fig. 1

Depositional History of Mississippian Sediments in Williston Basin

Accumulation of Mississippian oil in this area is strongly influenced by stratigraphy, and it is imperative to have an accurate understanding of the character and distribution of the formational subdivisions of the Mississippian system. A schematic table of Mississippian formations in the area is shown in Figure 3. The division of the Mississippian into Lodgepole, Mission Canyon, and Charles formations is as much a controversial issue as ever, if one judges by current literature.

An analysis of carefully selected Mississippian time lithological units will demonstrate persistent

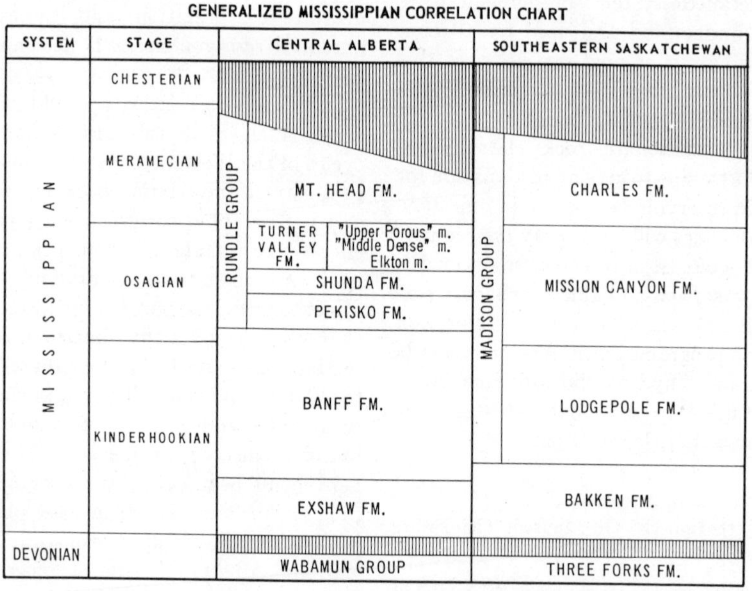

GENERALIZED MISSISSIPPIAN CORRELATION CHART
SYSTEM
STAGE
CENTRAL ALBERTA
SOUTHEASTERN SASKATCHEWAN
MISSISSIPPIAN
CHESTERIAN
MERAMECIAN
OSAGIAN
KINDERHOOKIAN
RUNDLE GROUP
MT. HEAD FM.
TURNER VALLEY FM.
"Upper Porous" m.
"Middle Dense" m.
Elkton m.
SHUNDA FM.
PEKISKO FM.
BANFF FM.
EXSHAW FM.
MADISON GROUP
CHARLES FM.
MISSION CANYON FM.
LODGEPOLE FM.
BAKKEN FM.
DEVONIAN
WABAMUN GROUP
THREE FORKS FM.

Fig. 2

basin, stable shelf, and unstable shelf relationships over the area. In general, there is a clear relation between the gross facies of these units and known oil occurrences. Evaporites in the Mississippian carbonate sequence definitely reflect the tectonic conditions under which they were deposited, and have been divided into intra-basinal and basin-margin occurrences. The separation of these two evaporitic environments was deemed so important for regional stratigraphic analysis that it was found necessary to redefine the lower limits of the Charles formation at its type locality.

Correlation of Mississippian rocks in the Williston basin is shown in Figure 3.

The evolution of the Mississippian depositional patterns in the Williston basin bears a remarkable resemblance to the sequence of West Texas-New Mexico Permian sedimentation as summarized by L. L. Sloss (1953).

The basic pattern of successive environmental and tectonic influences on sedimentation suggests the following sequence.

1. Establishment of a fully differentiated tectonic framework in the earliest Mississippian (Kinderhookian) with restricted, euxinic environments in the basins (formation of black shales and bituminous, micrograined limestones), normal marine conditions on the surrounding shelves, and brackish and terrestrial deposition at the margins of the sea.

2. Continuation of the initial pattern into the Osagian but with the establishment of saline (evaporitic) environments at the distal margins of the sea. Maximum carbonate development on the shelves as a result of an over-all, basinward migration of a shallow epeiric sea.

3. Formation of a single, markedly restricted evaporitic basin (Meramecian). Concentration of halite along the negative axis of the basin.

4. Tectonic stabilization of the entire area and deposition of a veneer of terrestrial redbeds (basal Chesterian).

The transition from marine limestone into marginal evaporites in the Mission Canyon formation has led to widespread confusion and miscorrelation. In a paper written in 1954, the senior writer advocated the use of persistent silt, sand, and argillaceous carbonate horizons as time boundaries in subdividing a carbonate, evaporite sequence in the northeastern parts of the Williston basin. These silt and mud incursions which terminate thin, carbonate depositional cycles were the products of epeirogenic fluctuations of the area. In that early paper, as a result of lack of well control, incomplete knowledge of the type Mississippian sections, and bad judgment on the writer's part, only five members of the Mission Canyon formation were recognized. However, even at that early stage, the "Midale" and overlying carbonate cycle were placed within the Mission Canyon formation. With increased well control it soon became obvious that at least nine cartographic units could have been used. Industry now includes the "Midale" (M.C. 7) pay in the Charles formation, although it is evident that this unit is genetically related to the interplay of transgression and regression of an over-all, retreating Mission Canyon sea (Fig. 3).

With this broad introduction to Mississippian sedimentation in the Williston basin, some highlights of the facies and porosity relations of the "Midale" carbonate cycle on the Souris Valley shelf feature of southeastern Saskatchewan are now presented.

TEXTURAL AND RESERVOIR PROPERTIES OF "MIDALE" CARBONATES (MISSION CANYON FORMATION) IN KINGSFORD-FLORENCE PRODUCING AREAS OF SOUTH-EASTERN SASKATCHEWAN

Large stratigraphic oil pools have been discovered at or near the Paleozoic subcrop of the "Midale" carbonate cycle in southeastern Saskatchewan (Fig. 4a). However, it was found that, though mechanical logs indicated fairly constant porosity up to the subcrop area, the character and distribution of carbonate lithosomes in the cycle controls the quality of production in these pools. There is a wealth of core information along this trend, because reservoir evaluation of the different pools is accomplished mainly through an extensive coring program.

Figure 4b, a section across the East Kingsford-Steelman oil fields, illustrates some of the facies changes of this depositional cycle toward the Paleozoic subcrop. Attempts have been made by some workers to use the marginal evaporite in the cycle as a time-stratigraphic unit, but local patch or shoal reefs of porous algal limestone were being deposited amidst submerged shoals of calcareous fossil debris and oölites at the same time when evaporites were being precipitated in back lagoonal areas. A thin, chalky to earthy, locally silty, argillaceous dolomite bed (M.C. 6) which terminates the M.C. 5 carbonate cycle has been used by the writers to delineate the base of the "Midale" cycle. This bed retains its lithological and electric-log characteristics over the Souris Valley platform.

The "Midale" carbonates are overlain by an-

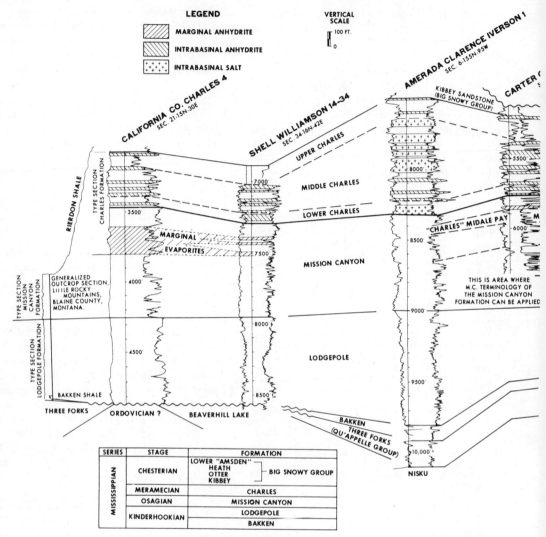

FIG. 3

hydrite and underlain throughout most of the area by an anhydrite floor. At the Paleozoic subcrop there is considerable secondary anhydrite infilling of primary carbonate porosity accompanied by dolomitization. The topographic highs on the Paleozoic erosional surface in parts of the Florence field, for example, are locally much brecciated, veined with gray or white anhydrite, and contain inclusions of the overlying Watrous redbeds. Completely dolomitized sections at the subcrop are very interesting because of the preservation of relict fossil structures.

Following the termination of deposition of the

M.C. 5 carbonate cycle by a veneer of chalky to earthy argillaceous dolomite, continued withdrawal of a shallow sea over the Souris Valley shelf resulted in the development of relict shoreline areas in which evaporites were deposited. Small patch or shoal reefs of vuggy, algal encrusted, bar-like material are found in juxtaposition to the marginal anhydrite sheet in the "Midale" beds (Kingsford area). These reefs could have acted as minor silling features necessary for the restriction and subsequent evaporation of the back lagoonal areas.

Calcareous algae played an important role in

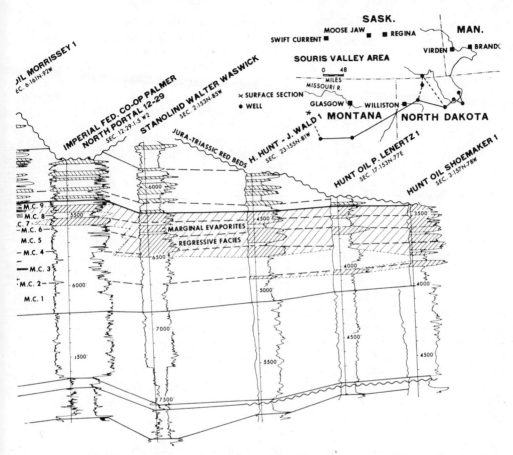

CORRELATION OF MISSISSIPPIAN ROCKS IN WILLISTON BASIN
LITTLE ROCKY MTNS. MONTANA - HUNT SHOEMAKER 1 N. DAKOTA
DECEMBER 1955

limestone building on the interior of the Souris Valley shelf area of southeastern Saskatchewan. J. Harlan Johnson (1956) has described in detail several genera of calcareous algae from thin sections of Saskatchewan material sent to him by one of the writers. Some stages in stabilization and encrustation of submerged, drifting shoals of calcareous, skeletal, and non-skeletal debris to form vuggy, algal, reef-like bodies are shown in Plate 1. Some of the strandline deposits have the appearance of oölite sand bars which have been stabilized by sediment-binding, encrusting calcareous algae. According to published reports, blue-green

algae are commonly the earliest colonizers of sediment newly deposited on tidal flats. It is well known that such algae exert a strong stabilizing effect upon the sediment they colonize, largely because they bind it together with their growing filaments and ultimately cover it with a mass of felted tubes. The stabilization of drifting sand-like material is really incipient reef growth. Photochemical removal of carbon dioxide from sea water by sea plants causes a decrease in bicarbonate ions and thus promotes the precipitation of calcium carbonate. Calcium carbonate is precipitated as a colloidal gel encrusting the leaves and

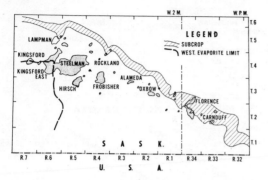

<figure>Fig. 4a.—"Midale" oil fields.</figure>

stems of these plants. The end-result is the production of cryptograined limestone which contains "syneresis" cracks and associated primary contraction vugs. The algal "knolls" of the Kingsford area generally have excellent horizontal as opposed to vertical permeability because of thin, cryptograined layers of encrusting algal limestone which separates partly encrusted, vuggy, pisolitic, and oölitic sections. From a reservoir point of view these carbonates are difficult to analyze because of anhydrite and calcite infilling of vugs and irregular, algal carbonate mud encrustation of original intergranular porosity. One has to resort to recording the percentage of effective void space.

Screened lenses of porous, finely comminuted carbonates are found draped over the wave resistant algal "knolls" of the West and East Kingsford area (Fig. 4b). Porosity, permeability, and oil saturation properties of these carbonates could be directly related to quantitative, textural (including grain size) carbonate measurements. The chart of Figure 5 shows that carbonate-matrix textural studies were the key to effective reservoir distribution in the upper "Midale" carbonates. Most of these carbonates consist of skeletal and oölitic limestones which have a finely comminuted, commonly dolomitized, limestone matrix with intergranular and chalky porosity. Effective reservoir porosity is controlled by the relative distribution and grain size of this matrix. Similar matrix grain size and effective reservoir relations have been found in the Triassic Boundary Lake lithosome in British Columbia and the Swan Hills member of the Beaverhill Lake formation in Alberta.

Bar-like trends of frequently current-bedded, skeletal, or non-skeletal material of very fine silt dimensions can be mapped in the Steelman and Oxbow areas of southeastern Saskatchewan. These deposits are usually extremely well sorted and are thought to have been developed by the attrition of crinoidal, algal, and oölitic material in

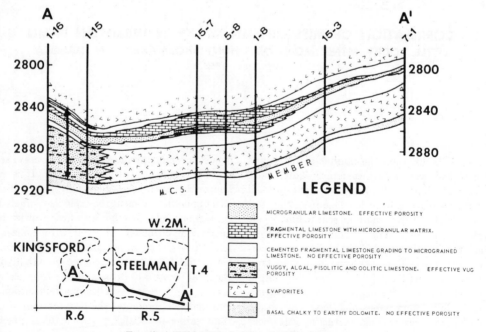

LEGEND

MICROGRANULAR LIMESTONE. EFFECTIVE POROSITY

FRAGMENTAL LIMESTONE WITH MICROGRANULAR MATRIX. EFFECTIVE POROSITY

CEMENTED FRAGMENTAL LIMESTONE GRADING TO MICROGRAINED LIMESTONE. NO EFFECTIVE POROSITY

VUGGY, ALGAL, PISOLITIC AND OOLITIC LIMESTONE. EFFECTIVE VUG POROSITY

EVAPORITES

BASAL CHALKY TO EARTHY DOLOMITE. NO EFFECTIVE POROSITY

Fig. 4b.—"Midale" beds, Kingsford-Steelman area.

a.—Reefal algal limestone. Oölitic and pisolitic bodies encrusted and bound together by layers of originally gelatinous calcareous mud precipitated by algal action. Drying of this material has produced primary contraction vugs. Excellent reservoir rock with much greater horizontal than vertical permeability. Vertical core.

b.—Fine-grained, well sorted, oölitic, and accretionary composite lump limestone (Bahama sands type), partly encrusted by large algal, pisolitic, and "biscuit" bodies. Horizontal core.

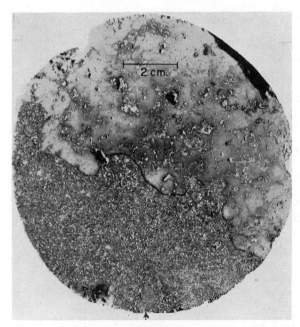

c.—Algal limestone with colloform mammillary surfaces and shrinkage cracks encrusting fine-grained, well sorted oölite and comminuted fossil debris. Some algal bodies in groundmass have been identified as Codiacean green algae. Horizontal core.

PLATE 1.—"Midale" carbonates.

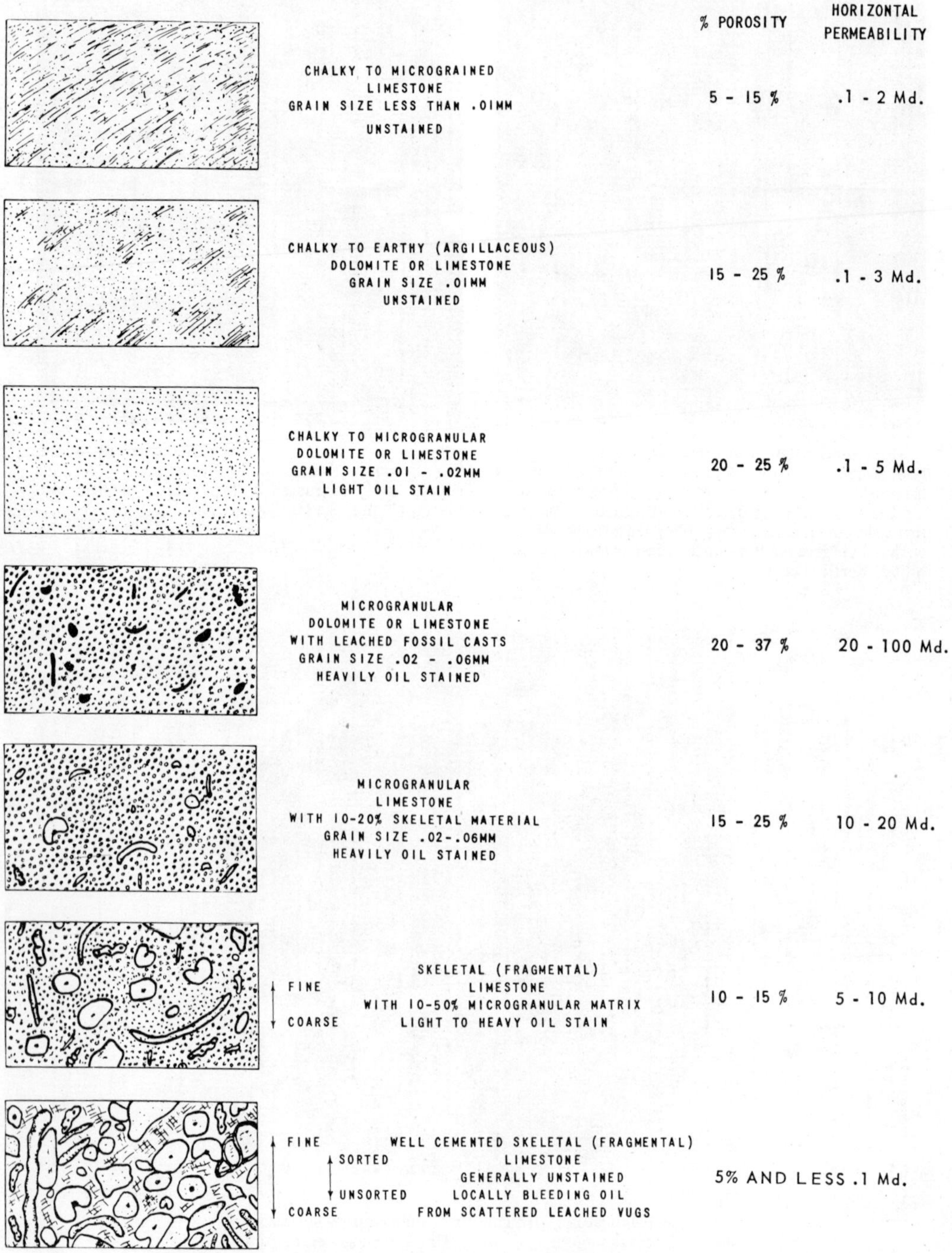

	% POROSITY	HORIZONTAL PERMEABILITY
CHALKY TO MICROGRAINED LIMESTONE GRAIN SIZE LESS THAN .01MM UNSTAINED	5 - 15 %	.1 - 2 Md.
CHALKY TO EARTHY (ARGILLACEOUS) DOLOMITE OR LIMESTONE GRAIN SIZE .01MM UNSTAINED	15 - 25 %	.1 - 3 Md.
CHALKY TO MICROGRANULAR DOLOMITE OR LIMESTONE GRAIN SIZE .01 - .02MM LIGHT OIL STAIN	20 - 25 %	.1 - 5 Md.
MICROGRANULAR DOLOMITE OR LIMESTONE WITH LEACHED FOSSIL CASTS GRAIN SIZE .02 - .06MM HEAVILY OIL STAINED	20 - 37 %	20 - 100 Md.
MICROGRANULAR LIMESTONE WITH 10-20% SKELETAL MATERIAL GRAIN SIZE .02-.06MM HEAVILY OIL STAINED	15 - 25 %	10 - 20 Md.
SKELETAL (FRAGMENTAL) LIMESTONE WITH 10-50% MICROGRANULAR MATRIX LIGHT TO HEAVY OIL STAIN (FINE / COARSE)	10 - 15 %	5 - 10 Md.
WELL CEMENTED SKELETAL (FRAGMENTAL) LIMESTONE GENERALLY UNSTAINED LOCALLY BLEEDING OIL FROM SCATTERED LEACHED VUGS (FINE / SORTED / UNSORTED / COARSE)	5% AND LESS	.1 Md.

FIG. 5.—Relation of textural variations, "Midale" carbonates with oil-saturation and porosity-permeability determinations.

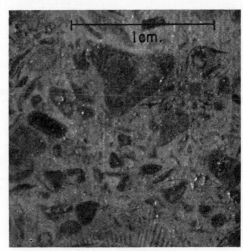

a.—Dolomitized microgranular limestone with leached fossil casts. 28% Porosity. 60 Millidarcys. Horizontal core.

b.—Skeletal limestone with microgranular matrix. 12% Porosity. 5 Millidarcys. Horizontal core.

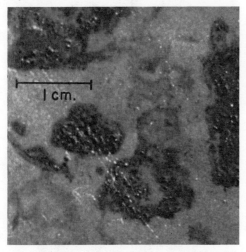

c.—Chalky to microgranular limestone with algal "cabbages" enclosing fine pellet material. Horizontal core.

d. Spiriferids in chalky to earthy dolomitic limestone. Horizontal core.

PLATE 2.—"Midale" carbonates.

current-agitated areas. Floating, calcareous, planktonic algae (Coccolithophoridae) possibly contributed to the formation of the microgranular material. Porosity in this class of carbonates is high (up to 37%, with permeabilities up to 120 millidarcys). The highest permeabilities occur where fossil (usually ostracod) casts supplement the pore space between the packed granules (Pl. 2a). This permeability is proportionately reduced when unleached skeletal material remains in the rock (Fig. 5 and Pl. 2b). The packed granules (usually in the 0.02–0.06 mm.-grain-size range) are particularly susceptible to dolomitization processes giving rise to a crystalline-granular texture.

All transitions to a rhombic (saccharoidal) dolomite with leached fossil casts can be seen particularly near the subcrop, although in this area the textures are partly masked by secondary anhydrite permeation. The pellicles of dolomite around the granules have a deterring effect on acid treatment results in the Kingsford-Steelman fields. In these fields the microgranular carbonate lenses are fairly continuous and should respond favorably to water injection processes (Fig. 4b).

These microgranular carbonates grade vertically and laterally into chalky to micrograined carbonates, which are partly of chemical origin but probably represent the "flour" formed by disintegration and abrasion of fossil debris and algal growths developed on the Souris Valley shelf. This "flour" or fine suspension material settled in the quiet-water environments of shelf-lagoons and around shoal areas during periods of relative quiescence in current activity. With decrease in grain size and increase in the amount of fine, chalky to clay-like material, the microtextured carbonates lose their oil-wetting ability and have high connate water saturation. The writers include the chalky to microgranular carbonates (grain size 0.01–0.02 mm.) in the effective reservoir material because they usually have some oil saturation. The chalky carbonate sections generally contain abundant ostracod remains. In the Florence-Oxbow areas scattered algal cabbages and fine pellet material are commonly observed (Pl. 2c). Well preserved brachiopods and articulate crinoid stems are usually found in the relatively thin beds of chalky to earthy, slightly argillaceous carbonates which can be traced over large areas of the shelf by their mechanical-log and lithological characteristics (Pl. 2d).

Lithographic, cryptograined to micrograined carbonates and skeletal to non-skeletal carbonates with a variable, lithified, carbonate-mud matrix are also well developed in the intershoal areas. All gradations from cryptograined limestone to skeletal or non-skeletal limestone with, for example, 5–10 per cent cryptograined matrix can be seen. All of this heterogeneous material appears to be genetically related and can be mapped and classed as belonging to the same group of carbonates. Even though this group locally contains a considerable amount of skeletal or non-skeletal grains, the sea currents during deposition were not strong enough or persistent enough to winnow away the carbonate ooze which remains as a matrix. Calcite cement is very subordinate or lacking simply because no pore space was available.

In the Alameda-Oxbow and Florence areas there is a thin development of algal lump and pellet limestone at the base of the "Midale" carbonate cycle. Plate 3a illustrates a relatively unce-

a.—Uncemented pellet limestone. 21% Porosity. 225 Millidarcys. Horizontal core.

b.—Brown anhydrite development at porosity permeability differences in "Midale" carbonates. Vertical core.

Description from top to bottom of core:

Inches
1   Chalky to microgranular dolomitic limestone (.01–.02 mm.). Light oil staining.
$\frac{1}{16}$ Brown anhydrite layer.
2   Chalky to earthy dolomite (.01 mm.). Scattered brown anhydrite crystals at base. No oil staining.
$\frac{1}{2}$ Brown anhydrite with carbonaceous laminae.
5   Cryptograined limestone with abundant vague pisolitic bodies partly replaced by brown anhydrite porphyroblasts.

PLATE 3.—"Midale" carbonates.

mented pellet limestone (Florence-Glen Ewen area) which was probably produced by the fragmentation of algal colonies. Thin stringers of this highly permeable pellet, locally oölitic material contribute mainly to the high initial flow potential of some of the Florence-Glen Ewen wells. This pellet material is associated with, and grades toward the west into, cryptograined limestones containing abundant algal "cabbages," lumps, and scattered leached ostracod casts. Due to the erratic nature of the pin-point vugs in the algal material and the development of leached ostracod casts, it is difficult to analyze the effectiveness of this type of limestone which has a bleeding type of oil staining. This basal, generally carbonate-mud rich unit could possibly have been laid down contemporaneously with the upper parts of the algal pisolite facies of East Kingsford.

Of particular interest are the brown porphyroblasts of "metasomatic," euhedral, anhydritic crystals, which are found scattered through the "Midale" carbonate sections. The crystals are brown because of hydrocarbon inclusions and commonly replace different types of carbonates. Oölites, for example, are commonly included in these brown anhydrites. Preliminary work suggested that since the "Midale" carbonates are sandwiched between primary anhydrite sheets over most of the area, a "wave" of metasomatism from the top anhydrite has contributed to the oil-trapping mechanism for the "Midale" cycle. The writers now think that even though some irregular anhydrite replacement of carbonates has taken place, this "metasomatism" has not materially altered their reservoir characteristics. Thin layers and scattered brown anhydrite crystals are usually found at porosity and permeability breaks within the "Midale" carbonate sections (Pl. 3b). This suggests that a salt-filtration mechanism affecting calcium sulphate-rich, circulating waters could be an explanation for this anhydrite formation. The distribution of these "metasomatic" anhydrite crystals bears no relation to the position of the overlying primary anhydrite or the basal, marginal anhydrite sheet of the "Midale" cycle.

When a clasticity index approach is applied to "Midale" carbonates, the final textural results can easily be contoured into linear patterns which appear to fit oil-production behavior of the various wells. The topographical expression of the algal banks west of the evaporite strandline must have had an important influence on current refraction patterns during later "Midale" carbonate deposition. These current patterns sifted the carbonates with effective matrix porosity into linear belts or shoals which run transverse to the Paleozoic subcrop trend of the "Midale" beds. The bar-like trends of carbonates with effective porosity are separated by quiet water, intershoal areas in which are found higher percentages of shelf-lagoonal carbonate muds, chalky carbonates, and fragmentals with a mud matrix, which have no effective porosity. With the use of effective porosity isopachs, isoporosity feet, and facies maps, one can be selective about development and wildcat acreage in an area formerly thought to be one huge stratigraphic pool.

TEXTURAL AND RESERVOIR PROPERTIES OF "ELKTON" CARBONATE CYCLE (TURNER VALLEY FORMATION) IN SOUTHWESTERN ALBERTA

Official nomenclature of well defined Upper Mississippian (Rundle group) shelf-carbonate cycles in southwestern Alberta is still burdened with old Turner Valley field names, such as "Upper Porous" and "Middle Dense" zones. Widespread transgressive sheets of coarse, generally dolomitized, fragmental (skeletal) limestones are separated by shallow-water depositional units of silty, locally cherty, lithified carbonate muds which can be established as time-stratigraphic boundaries for correlation purposes. However, it is not the purpose of this report to condemn present Mississippian nomenclature or to rename some of the carbonate cycles as has been done in southeastern Saskatchewan.

Major hydrocarbon (oil and gas) reserves have already been discovered in the Mississippian "Elkton" carbonate cycle, both in the Foothills Belt and along the subcrop in southwestern Alberta. The "Elkton" carbonate cycle generally consists of coarse skeletal (predominantly crinoidal) carbonates, ranging in thickness from 80 to 150 feet and of variable porosity and permeability. These carbonates are overlain and underlain by tight, lithified, silty carbonate mud deposits up to the subcrop area, where the eroded reservoir material is covered by generally impermeable Mesozoic shales and silty sandstones. Porosity and permeability properties of the producing intervals in the Sundre, Westward Ho, and Harmattan-Elkton fields, situated at or near the "Elkton" subcrop, can be directly related to quantitative, carbonate textural measurements. In this area, as opposed to the "Midale" cycle, mechanical logs (neutron, microlog, and micro-laterolog) can be used to differentiate effective and non-effective reservoir types. Effective reservoir material of this cycle was found to consist mainly of the dolomitized equivalent of an originally coarse skeletal limestone with a variable amount of porous, finely comminuted (granular) skeletal matrix. Primary porosity was very important in the control of dolomitization, which began with the replacement of this matrix by euhedral rhombohedrons and finally affected the coarse, skeletal material (now generally indicated by leached, fossil cast outlines). These porous dolomites grade laterally in a predictable way into tight, relatively non-dolomitized, well sorted, coarse skeletal limetones, with original high interfragmental porosity now completely infilled with clear crystalline calcite. This lithification by cementation took place early in the history of carbonate sedimentation of this area and before secondary dolomitization processes took effect. Present hydrocarbon accumulation along the subcrop is controlled largely by updip truncation of the "Elkton" member. However, it is also strongly influenced by primary porosity pinchouts caused by lateral facies changes from dolomitized, leached, skeletal limestones with matrix into tight, cemented skeletal limestones. Similar facies changes exist in the Turner Valley oil field, suggesting primary hydrocarbon accumulation in the "Elkton" member before the Laramide structural movements took place.

At the Jasper Conference of the American Association of Petroleum Geologists, in September, 1955, D. G. Penner introduced the name "Elkton" member for the lower bioclastic or skeletal rich unit of the Turner Valley formation. In 1957 the proposed nomenclature was further refined by Penner. The type section of this member was designated as that penetrated in the Great Plains *et al.* Elkton No. 16-13 well (Lsd. 16, Sec. 13, T. 31, R. 4, W. 5th Meridian) between the depths of 8,705 and 8,845 feet. Development drilling in the Harmattan-Elkton field showed that it was possible to subdivide the "Elkton" member into three sub-members (Penner, 1957). The "Elkton" member, 140 feet thick, was equated to the combined "Lower Porous" and "Crystalline Zone" of the Turner Valley field.

Penner's published cross section from Elkton No. 16-13 to the Pine Creek well (Lsd. 12, Sec. 12, T. 20, R. 2, W. 5th Meridian) and intervening locations is shown in the upper half of Figure 6. While commending Penner's clarification of the relation of the Sundre-Harmattan producing intervals to those of the Turner Valley oil field, for regional correlation purposes, the writers have had to redefine the upper limits of his "Elkton" member.

The "A" sub-member in the Harmattan-Elkton field was found to consist generally of a microcrystalline dolomite with scattered relict skeletal material and abundant inclusions of milky white chert. Rock photograph of Plate 4a is fairly typical of this sub-member and shows it to be the dolomitized equivalent of washes of micro-finely comminuted, skeletal material deposited in a carbonate mud environment. Chert nodules, quartz silt partings, and scattered plant fragment traces are common, suggesting muddy, shallow-water deposition which was unfavorable for much skeletal carbonate or good reservoir development. Microcrystalline (anhedral to subhedral) calcareous dolomites are generally the rule, with fairly high porosity (commonly more than 10%) and low permeability. These dolomites commonly give rise to substantial blows of gas on drill-stem test. Locally within the field, the "A" sub-member has good porosity and permeability, due to an increase in the amount of coarser-textured, more rhombic dolomite whose intercrystalline porosity has been supplemented by the presence of scattered, leached, coarse skeletal material. Within and away from the Harmattan-Elkton field these coarser-textured, rhombic dolomites with effective porosity grade laterally and vertically into cherty, silty, and argillaceous, microcrystalline dolomites that are texturally indistinguishable from Penner's middle sub-member.

On the basis of carbonate texture, chert and silt content, depositional environment, timestratigraphic correlation, and general reservoir properties, Penner's middle and upper "Elkton" sub-members are considered to be lateral equivalents of the cherty, "Middle-Hard" or "Middle-Dense" zone of the Turner Valley oil field. For convenience the writers have designated Penner's lower sub-member, or main prospective zone, as the "Elkton" carbonate cycle.

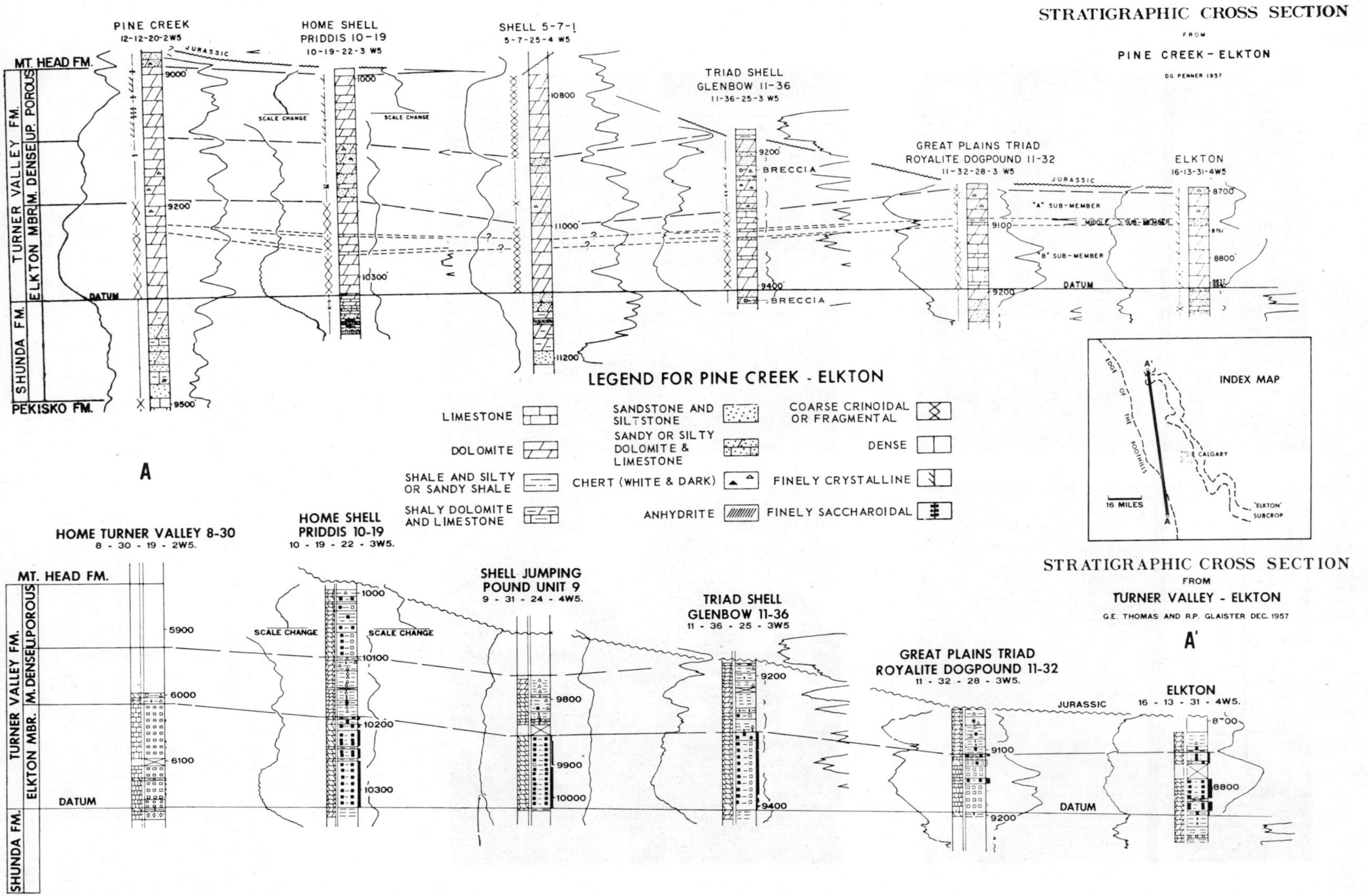

FIG. 6.—Stratigraphic cross sections.

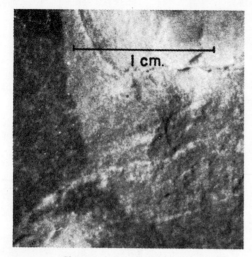

a.—Cherty microcrystalline dolomite.

b.—Cemented skeletal (crinoid-bryozoa) limestone.

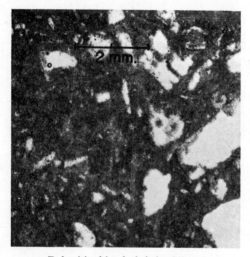

c.—Dolomitized leached skeletal limestone with porous matrix.

d.—Micro-rhombic dolomite with leached fossil casts.

PLATE 4.—"Elkton" carbonates.

"ELKTON" CARBONATES WITH NO EFFECTIVE POROSITY

For mapping purposes, "Elkton" carbonates can be broken down into effective and non-effective porosity units.

CEMENTED SKELETAL AND NON-SKELETAL CARBONATES

Of great interest is the widespread occurrence in the "Elkton" cycle of thick deposits of generally coarse, cemented skeletal limestones ("crystalline" limestones of former workers). These are composed predominantly of the calcareous remains of disarticulated echinoderm ossicles and plates. Although space between crinoid fragments in rocks of this type consists largely of clear crystalline calcite, few samples lack some organic remains of non-crinoidal origin. Most commonly the "foreign" material consists of bryozoan fragments. These tight, coarse, skeletal limestones, with high original interfragmental porosity, now completely infilled with clear crystalline calcite, usually analyze less than 5 per cent porosity and 0.1 MDS. Apparently, earlier workers did not recognize echinoderm plates and ossicles as rock-builders in the lower "crystalline" zone of the Turner Valley oil field and referred to them as calcite crystals. Even in recent excellent papers (e.g., Hemphill, 1957) occur statements such as "in the No. 9-22 well the section consisted of limestone, medium-grey, medium to coarsely crystalline, and slightly fossiliferous."

The thin-section photograph of Plate 4b should clarify concepts of what constitutes a "crystalline" or cemented skeletal limestone. Even though cemented skeletal or non-skeletal limestones are found from the Cambrian to Quaternary and contribute to most carbonate text-book photographs, extremely few authors have committed themselves in mode of origin discussions. The clear crystalline calcite cement of the sorted skeletal or non-skeletal limestones might be interpreted as reorganized carbonate mud matrix or as primary calcite cement. The writers concur with R. C. Moore's observations on the Mississippian of the Ozarks and believe it to be a primary chemical precipitate for the following reasons.

1. The crystalline cement is present in considerable amounts only in skeletal or non-skeletal limestones that have a relatively high degree of sorting and rounding. This suggests that much of the interstitial fine material was winnowed out by strong currents or shoaling water where wave action was effective. This winnowing process would create interstitial voids favorable for the formation of primary crystalline cement.

2. Prominent crystalline calcite overgrowths on many grains, especially on crinoid columnals, are in optical continuity. According to Pettijohn, this continuity is a characteristic of primary cement.

3. There are no relict structures of grains or comminuted fossils in the crystalline cement as would be expected if the cement was the product of a reorganized or replaced matrix. The cement was probably introduced into open pores in the course of diagenesis, being precipitated as crystal growths derived from carbonate-saturated waters.

4. Edges of the fossils or non-skeletal material are not corroded or altered in a manner suggesting effects of recrystallization.

This lithification by cementation took place early in the carbonate sedimentation history of this area, and before secondary dolomitization processes took effect. The very nature of the clear crystalline calcite cement infilling of primary interfragmental porosity inhibited dolomitization. Dolomitization of these limestones could develop along cleavage cracks in the calcite cement or along incipient fractures. Cemented skeletal or non-skeletal limestones in Devonian or Mississippian sections of the Rockies or Foothills Belt, usually show effects of dolomitization processes as a result of stresses induced by mountain-building.

Oölitic and associated surficially coated grains occur locally in cemented skeletal limestones of the "Elkton" member. Localities where these fringing shoal deposits have been found include the center part of the Turner Valley oil field, and the Blackie, Brant, Dogpound, and Sundre areas. For mapping purposes, these generally cemented, well sorted, oölitic limestones can be grouped with cemented skeletal limestones, with which they are closely associated. Oölites are rare in unsorted skeletal limestones.

SKELETAL OR NON-SKELETAL LIMESTONES WITH CHALKY TO MICROGRAINED MATRIX

Of local interest are the skeletal or non-skeletal limestones containing a chalky matrix which are found in the Brant and Blackie areas, south of Calgary. Poor grain-sorting, presence of delicate bryozoan fronds, and chalky to micrograined matrix, all suggest sheltered, quiet-water conditions of deposition. The chalky matrix material has high connate water content and virtually no oil saturation because of the fine capillary pores. These were the only areas where the microlog curve gave unreliable estimates of effective porosity in the "Elkton" reservoir.

DOLOMITIZED CARBONATE MUDS AND NON-LEACHED SKELETAL TO NON-SKELETAL CARBONATES WITH TIGHT, ORIGINALLY CARBONATE MUD MATRIX

All gradations from original cryptograined limestone to relict skeletal or non-skeletal limestone (e.g., with 5–10% cryptograined matrix) can be seen. The carbonate mud was apparently easily dolomitized to produce a cryptocrystalline to microcrystalline, anhedral, interlocking type of dolomite with no effective porosity.

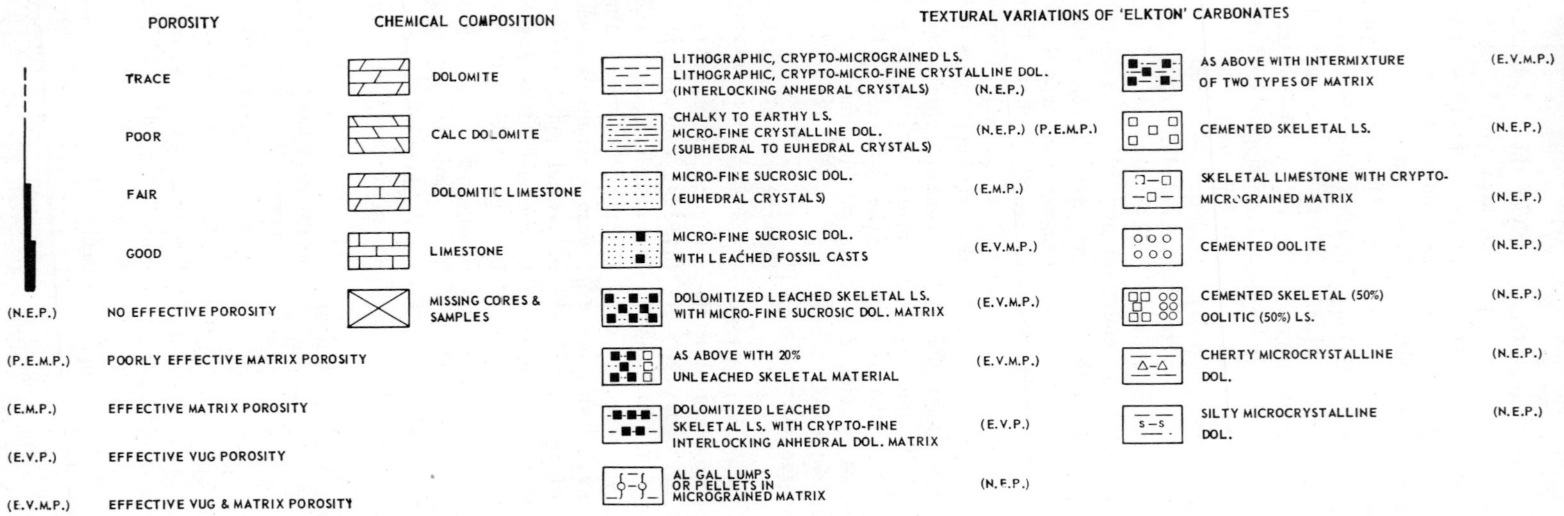

POROSITY

TRACE
POOR
FAIR
GOOD

NO EFFECTIVE POROSITY (N.E.P.)
POORLY EFFECTIVE MATRIX POROSITY (P.E.M.P.)
EFFECTIVE MATRIX POROSITY (E.M.P.)
EFFECTIVE VUG POROSITY (E.V.P.)
EFFECTIVE VUG & MATRIX POROSITY (E.V.M.P.)

CHEMICAL COMPOSITION

DOLOMITE
CALC DOLOMITE
DOLOMITIC LIMESTONE
LIMESTONE
MISSING CORES & SAMPLES

TEXTURAL VARIATIONS OF 'ELKTON' CARBONATES

LITHOGRAPHIC, CRYPTO-MICROGRAINED LS. LITHOGRAPHIC, CRYPTO-MICRO-FINE CRYSTALLINE DOL. (INTERLOCKING ANHEDRAL CRYSTALS) (N.E.P.)

CHALKY TO EARTHY LS. MICRO-FINE CRYSTALLINE DOL. (SUBHEDRAL TO EUHEDRAL CRYSTALS) (N.E.P.) (P.E.M.P.)

MICRO-FINE SUCROSIC DOL. (EUHEDRAL CRYSTALS) (E.M.P.)

MICRO-FINE SUCROSIC DOL. WITH LEACHED FOSSIL CASTS (E.V.M.P.)

DOLOMITIZED LEACHED SKELETAL LS. WITH MICRO-FINE SUCROSIC DOL. MATRIX (E.V.M.P.)

AS ABOVE WITH 20% UNLEACHED SKELETAL MATERIAL (E.V.M.P.)

DOLOMITIZED LEACHED SKELETAL LS. WITH CRYPTO-FINE INTERLOCKING ANHEDRAL DOL. MATRIX (E.V.P.)

ALGAL LUMPS OR PELLETS IN MICROGRAINED MATRIX (N.F.P.)

AS ABOVE WITH INTERMIXTURE OF TWO TYPES OF MATRIX (E.V.M.P.)

CEMENTED SKELETAL LS. (N.E.P.)

SKELETAL LIMESTONE WITH CRYPTO-MICRO-GRAINED MATRIX (N.E.P.)

CEMENTED OOLITE (N.E.P.)

CEMENTED SKELETAL (50%) OOLITIC (50%) LS. (N.E.P.)

CHERTY MICROCRYSTALLINE DOL. (N.E.P.)

SILTY MICROCRYSTALLINE DOL. (N.E.P.)

FIG. 7.—Legend for Figures 6, 8, and 9.

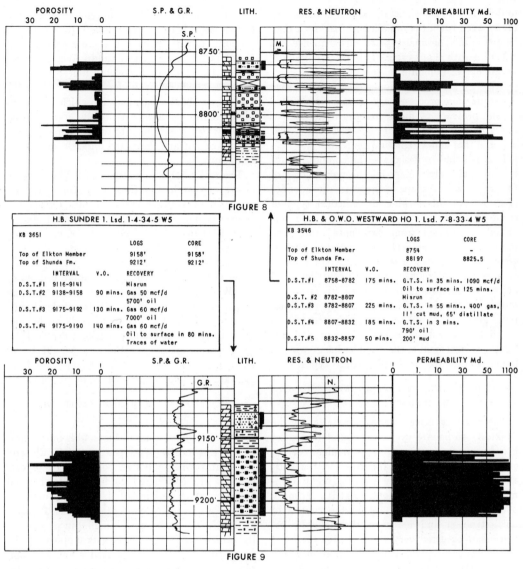

FIGURE 8

H.B. SUNDRE 1. Lsd. 1-4-34-5 W5

KB 3651

	LOGS	CORE
Top of Elkton Member	9158'	9158'
Top of Shunda Fm.	9212'	9212'

INTERVAL	V.O.	RECOVERY
D.S.T.#1 9116-9141		Misrun
D.S.T.#2 9138-9158	90 mins.	Gas 50 mcf/d 5700' oil
D.S.T.#3 9175-9192	130 mins.	Gas 60 mcf/d 7000' oil
D.S.T.#4 9175-9190	140 mins.	Gas 60 mcf/d Oil to surface in 80 mins. Traces of water

H.B. & O.W.O. WESTWARD HO 1. Lsd. 7-8-33-4 W5

KB 3546

	LOGS	CORE
Top of Elkton Member	8754	-
Top of Shunda Fm.	8819?	8825.5

INTERVAL	V.O.	RECOVERY
D.S.T.#1 8758-8782	175 mins.	G.T.S. in 35 mins. 1090 mcf/d Oil to surface in 125 mins.
D.S.T.#2 8782-8807		Misrun
D.S.T.#3 8782-8807	225 mins.	G.T.S. in 55 mins., 400' gas, 11' cut mud, 65' distillate
D.S.T.#4 8807-8832	185 mins.	G.T.S. in 3 mins. 790' oil
D.S.T.#5 8832-8857	50 mins.	200' mud

FIGURE 9

FIGS. 8–9

"ELKTON" CARBONATES WITH EFFECTIVE POROSITY

The only extensively developed reservoir rock in the "Elkton" member is dolomite. Investigations so far completed suggest that secondary dolomitization of skeletal and other limestones took place on a volume-for-volume relationship, and that the porosity of the resultant dolomite (apart from leaching effects) was inherited from the original limestone. The secondary dolomites are generally coarse-grained, many of them with relict limestone textures or casts of fossil debris.

On the basis of relict textures in these dolomites, it is possible to carry the zonation of limestone textural types into predominantly dolomite sections.

With regard to the relation of dolomite development to textural features of original limestones, it has been observed that it preferentially occurs in open pores or in matrix (chalky, granular, and carbonate mud) material that surrounds the larger skeletal or non-skeletal grains. These larger grains are generally the last to show conversion to dolomite. Many skeletal fragments remain as calcite even when the remainder of the rock may be dolomite. The final type in this sequence is a dolomite with fossil casts. This preferential development of dolomite in certain textural components of the original limestone suggests that dolomitization processes are strongly controlled by the presence of fluids in interfragmental or intergranular porosity or by carbonate mud material that had a high fluid content.

It is possible to designate the composition of the original matrix material through studies of the grain-size and shape of the resultant dolomite (Fig. 7). An interlocking or anhedral type of crypto-microcrystalline dolomite matrix is interpreted as derived from carbonate mud. The comminuted or pulverized, generally porous, granular and chalky material, either of skeletal or non-skeletal origin, commonly contributes to matrix or intergranular porosity in carbonate reservoirs. On dolomitization, this porous material gives rise to subhedral or euhedral (rhombic) dolomites with intercrystalline porosity, unless the enlargement of the granules has continued too far to produce "mosaic" textural types.

DOLOMITIZED LEACHED SKELETAL LIMESTONES WITH POROUS MATRIX

The most effective reservoir material of the "Elkton" carbonate cycle was found to consist mainly of the dolomitized equivalent of an originally coarse, skeletal limestone with a variable amount of generally porous, finely comminuted (granular) skeletal matrix. Primary porosity was very important in the control of dolomitization and much of the dolomite replacement occurred very shortly after the limestone was laid down. Dolomitization probably began with the replacement of the porous granular matrix by sub-euhedral rhombohedrons and finally affected the coarse, skeletal grain material (now generally indicated by leached fossil cast outlines). Highest permeabilities occur where fossil casts supplement the pore space between the packed dolomite rhombs (Pl. 4c). This permeability is proportionately reduced when relict unleached skeletal material remains in the rock.

The highest porosities and permeabilities are found in the relatively poorly sorted, dolomitized, leached skeletal limestones with porous matrix (up to 30% porosity and 1,000+ MDS.) These porous dolomites grade laterally in a predictable

way into tight, well sorted, cemented, skeletal limestones. Figures 8 and 9 illustrate that porosity logs readily distinguish these markedly different facies types in the Sundre-Westward Ho fields. The cemented skeletal limestones were probably laid down under initial high wave energy or shoal conditions in which the comminuted, microgranular to finely granular material, which contributes to matrix porosity, was winnowed out and deposited under lower energy conditions.

The rock photograph sequence of Plate 5 demonstrates that locally there is a progressive destruction of skeletal grain outlines during the process of dolomitization. The end-result of such dolomitization is the production of medium to coarse crystalline dolomite, the generic implications of which are in doubt. This type of material is common in the stratigraphic column of the Western Canada basin (e.g., the Devonian Nisku formation). When definite skeletal outlines have been destroyed and replaced by medium to coarse dolomite crystals, a crystallinity (dolomite grain size) ratio map would be of great value in differentiating dolomitized fragmentals and dolomitized carbonate mud areas.

MICRO TO FINE RHOMBIC DOLOMITES WITH LEACHED FOSSIL CASTS

The rock photograph of Plate 4d is representative of this locally developed group of "Elkton" carbonates with effective porosity. These carbonates are considered to be the dolomitized equivalents of comminuted, skeletal, or non-skeletal grains in current-agitated areas. Porosity in this class of carbonates is high, but permeability is high only where fossil casts supplement the pore space between the packed granules. Complex intermixtures of this group of carbonates with effective matrix and vug porosity and chalky to earthy carbonates with no effective porosity are usually found at the top of the "Elkton" member.

DOLOMITIZED CARBONATE MUDS WITH LEACHED FOSSIL CASTS AND LEACHED SKELETAL TO NON-SKELETAL CARBONATES WITH TIGHT, ORIGINALLY CARBONATE MUD MATRIX

Carbonate muds, probably because of an original high fluid content, alter easily to a crypto-microcrystalline, anhedral to subhedral, interlocking type of dolomite with little or no effective

a.—Dolomitized unleached skeletal limestone with porous matrix.

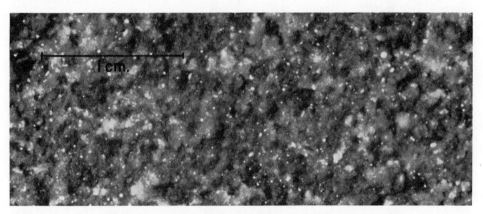

b.—Dolomitized partly leached skeletal limestone with porous matrix.

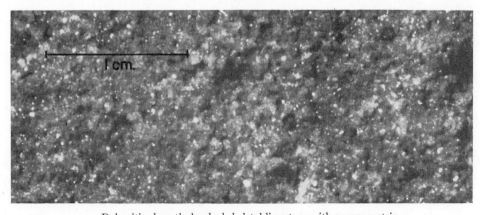

c.—Dolomitized partly leached skeletal limestone with porous matrix.
(Medium to coarse sub-euhedral dolomite)

PLATE 5.—"Elkton" carbonates. Demonstration series illustrating progressive destruction
of skeletal grain outlines during dolomitization.

porosity. However, the skeletal or non-skeletal grains embedded in this original carbonate mud material are locally leached to produce generally poorly effective vug porosity.

Conclusions

It is hoped that these observations on the relations of grain, matrix, and cement variants of carbonate rocks to porosity and permeability determinations will be helpful in the recognition and mapping of such associations in other places.

With regard to exploration philosophy on unconformity, "porosity-wedge" plays, these studies again reveal the necessity for reconstruction of the sedimentation history of the prospective, eroded unit. The mapping of dissected, primary permeability barriers (cemented skeletal limestones) at or near the "Elkton" subcrop is just as important to oil and gas exploration as is the recognition of impermeable, marginal anhydrites at or near the subcrop of the Mississippian shelf carbonate cycles in the Souris Valley area of southeastern Saskatchewan.

References

HEMPHILL, C. R., 1957, "History and Development of the Sundre, Westward Ho and Harmattan Oil Fields," *Jour. Alberta Soc. Petrol. Geol.*, Vol. 5, No. 10, pp. 232–47.

JOHNSON, J. H., 1956, "Studies on Mississippian Algae," *Quar. Colorado School Mines*, Vol. 54, No. 4 (October).

MOORE, R. C., 1957, "Mississippian Carbonate Deposits of the Ozark Region," *Soc. Econ. Paleon. and Mineral. Spec. Pub. 5.*

PENNER, D. G., 1957, "The Elkton Member," *Jour. Alberta Soc. Petrol. Geol.*, Vol. 5, No. 5 (May), pp. 101–04.

SLOSS, L. L., 1953, "The Significance of Evaporites," *Jour. Sed. Petrology*, Vol. 23, No. 3 (September), pp. 143–61.

THOMAS, G. E., 1954, "The Mississippian of the Northeastern Williston Basin," *Canadian Inst. Min. Met. Bull. 503* (March), pp. 136–42.

BULLETIN OF THE AMERICAN ASSOCIATION OF PETROLEUM GEOLOGISTS
VOL. 50, NO. 10 (OCTOBER, 1966), P. 2260-2268, 11 FIGS., 1 TABLE

REEF TRENDS OF MISSISSIPPIAN RATCLIFFE ZONE, NORTHEAST MONTANA AND NORTHWEST NORTH DAKOTA[1]

ALAN R. HANSEN[2]
Billings, Montana

ABSTRACT

Algal pelletoid reef deposits comprise the main oil-productive lithologic type in the Mississippian Ratcliffe zone of the Charles Formation, northeast Montana and northwest North Dakota. These rocks were formed in a moderate- to high-energy environment near the central part of the Williston basin.

The reef deposits appear to have formed in slightly shallower parts of the shallow Ratcliffe sea. These low-relief submarine highs were controlled partly by relatively small movements along fault trends and lines of structural weakness that were active intermittently during Paleozoic time.

INTRODUCTION

Within the 3,800-sq.-mi. area of Ratcliffe porosity development outlined on Figure 1, there are 12 oil fields which produce from the following stratigraphic units: the Mississippian Ratcliffe, 7 fields; the Devonian Nisku, 1 field; the Devonian Winnipegosis, 2 fields; the Silurian Interlake, 1 field; and the Ordovician Red River, 1 field. These 12 fields produced approximately 375,000 bbls. of oil during January, 1966.

The area outlined on Figure 1 has been part of the Williston basin from Early Ordovician time to the present, or about 500 million years. Conditions have been favorable for oil accumulation for nearly half of these 500 million years. This paper is concerned with the 250 million-year-old Ratcliffe limestone beds that are 6,500–7,200 ft. deep. These beds were formed in a relatively shallow sea which contained abundant organic life. Algal reefs grew and were buried, forming the host rock for indigenous and migrated oils. Based on the writer's estimates, less than half of the oil reserves in the Ratcliffe zone have been found.

Oil production from the Ratcliffe zone has been found in the fields tabulated on Table I. The discoveries of these fields were based on stratigraphic, structural, and isopachous studies, but were aided greatly by seismic interpretations, good fortune, and fortitude.

[1] Read before the Rocky Mountain Section of the Association at Billings, Montana, September 28, 1965. Manuscript received, February 28, 1966; accepted, June 2, 1966.

[2] District exploration geologist, Sun Oil Company. The writer appreciates the permission of Sun Oil Company to publish this paper. Special thanks are due Mathew S. Tudor and Wm. B. Dodds for suggestions and constructive criticism.

Rocks of the Ratcliffe zone are of Mississippian, early Meramecian, age as shown on the correlation chart (Fig. 2), and constitute a mappable unit of the Charles Formation throughout the subject area. The Ratcliffe zone is partly equivalent to the "C" zone, a productive interval in the Poplar field of Roosevelt County, Montana.

The term Ratcliffe was first used in this area as a field name in Saskatchewan, Canada (sec. 30 T. 1 N., R. 15 W-2). Ratcliffe field is a two-well field, discovered in 1952, and is still producing from the lower Charles Ratcliffe zone as designated in this paper.

STRUCTURE

Structural relief at the top of the Ratcliffe zone is shown on Figure 3. The Ratcliffe productive areas, magnetic positive trends, and surface alignments are superimposed on this structure map. More detailed mapping, not included here, shows pronounced northeast strike lines that coincide approximately with a regional pattern of faulting, topographic alignment of streams, and glacial lakes. Another significant trend is the northwest alignments of magnetic positives, seismic trends, and lower Paleozoic features as defined by isopachous studies. Much of the Ratcliffe zone production occurs in areas where these two alignments intersect.

Faulting.—Of particular interest is the fault and fracture trend that passes through the Dwyer-Grenora producing areas. This major line of weakness is interpreted to be a northeastern extension of the Weldon fault. This fault has a down-to-the-south surface displacement as great as 160 ft. in the Tertiary Fort Union beds 80 mi. southwest of the Dwyer area (Collier and Knechtel, 1939, p. 17). Norwood (1965), after

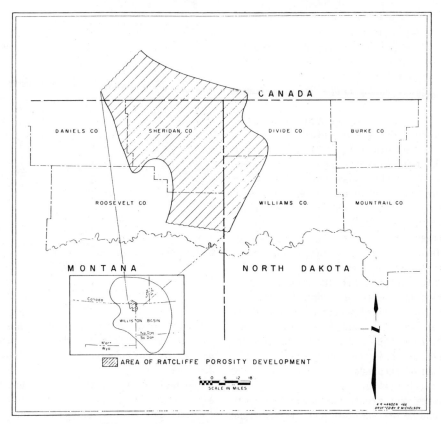

Fig. 1.—Index map, northwest North Dakota, northeast Montana, and adjacent Saskatchewan. Area of Mississippian Ratcliffe porosity development shown by diagonal lines.

studying the structures and faulting farther southwest in central Montana, believes that the Weldon fault and its indicated extension into the area of interest have been active intermittently since early Paleozoic time (E. E. Norwood, 1966, oral commun.). Additional evidence for this fault trend has been obtained from seismic-reflections in the Devonian near the Dwyer field. These data indicate faulting within the Devonian, the sense of displacement being down-to-the-south. This

TABLE I. OIL PRODUCTION BY FIELDS, MISSISSIPPIAN RATCLIFFE ZONE

State or Province Field Name	Discovered (Mo./Yr.)	Number of Wells	Production (Bbls.), January, 1966*	Cumulative (Bbls.) through December, 1965*
Montana				
Dwyer	9/60	15	30,599	3,146,116
Lone Tree	9/62	3	5,183	195,716
Goose Lake	5/63	19	67,709	1,950,738
Flat Lake	6/64	36	101,710	930,389
Shotgun Creek	9/64	1	—	40,059
North Dakota				
Grenora	2/61	15	21,324	1,593,737
Saskatchewan				
Flat Lake	6/64	13	46,164	95,829
Ratcliffe	11/52	2	710	114,878

* Current (May, 1966) oil production from the fields listed is approximately 274,000 bbls. per month.

provides evidence for at least pre-Mississippian faulting in the Dwyer area.

M. S. Tudor (1966, oral communication) has noted a consistent thickening of Mississippian and Pennsylvanian sediments north of the indicated northeast-trending fault. This thickening is interpreted by Tudor to be the result of vertical movement on an inferred pre-Ratcliffe-age fault, the displacement being down-to-the-north.

It seems apparent that, if the stresses that caused the initial movement along this fault trend were oriented similarly and were active periodically since early Paleozoic time, easily measurable faulting would be evident. However, in the Ratcliffe zone, it is difficult to measure any vertical displacement by log correlation even though much fracturing, some slickensides, and minor offsets were noted in cores from the Dwyer field.

It is believed likely that the stresses causing the relative movement along this particular fault have not been oriented similarly throughout the geologic history of the fault. Such conditions undoubtedly have made it difficult to recognize the fault and the consequent implications for reef growth and oil accumulation. It should be noted further that, because of the shallow nature of the Ratcliffe sea, only a very small amount of movement would be required to provide the higher-energy conditions favorable for reef development.

STRATIGRAPHY

Drilling to date has confirmed the existence of three major areas of Ratcliffe reef development and associated production: the Lone Tree-Flat Lake area along the Canadian border, the Goose Lake area 9 mi. south, and the Dwyer-Grenora area still farther south.

Figure 4 shows that at each of these producing areas the reservoir rock is primarily an algal pelletoid limestone. This facies first grades laterally into a "sub-chalky" facies, and next, in an off-reef direction, into a normal-marine sequence of interbedded fossiliferous and argillaceous limestone. Some shaly lagoonal deposits occur along the reef flanks in areas where water circulation was somewhat restricted.

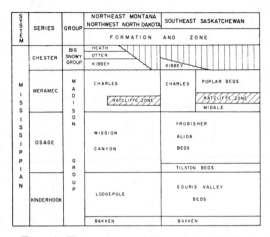

FIG. 2.—Time-rock correlation chart, northwest North Dakota, northeast Montana, and southeast Saskatchewan, showing stratigraphic position of Ratcliffe zone.

REEF DEVELOPMENT

The question arises regarding why these particular reefs grew where they did. This question is particularly pertinent in view of the once common suggestion that this entire area was a very stable part of the Williston basin during early Meramecian time. Prior to the Ratcliffe discoveries, the data on the Meramecian from scattered well control indicated that an open-marine environment had prevailed; this conclusion was based on the presence of similar suites of fossils and rocks throughout the known part of the area. No major interruptions within the known stratigraphic sequence were postulated and the chance that one or more profitable oil fields might occur locally seemed remote.

The three main productive areas are located along old Paleozoic positive trends, or lines of weakness, whose age and orientation have been determined by seismic, magnetic, and isopachous studies. The probable reason for the local reef development is that subsequent tectonic activity along these lines of weakness produced local positive areas in the Ratcliffe sea where algal reefs could develop.

This mild but persistent orogenic activity has

⟫⟫→

FIG. 3.—Structural contour map; structural datum is top of Ratcliffe zone. Contours in feet; elevation datum is sea-level. Map shows magnetic positive trends, inferred faults, geomorphic alignments, and Ratcliffe productive areas.

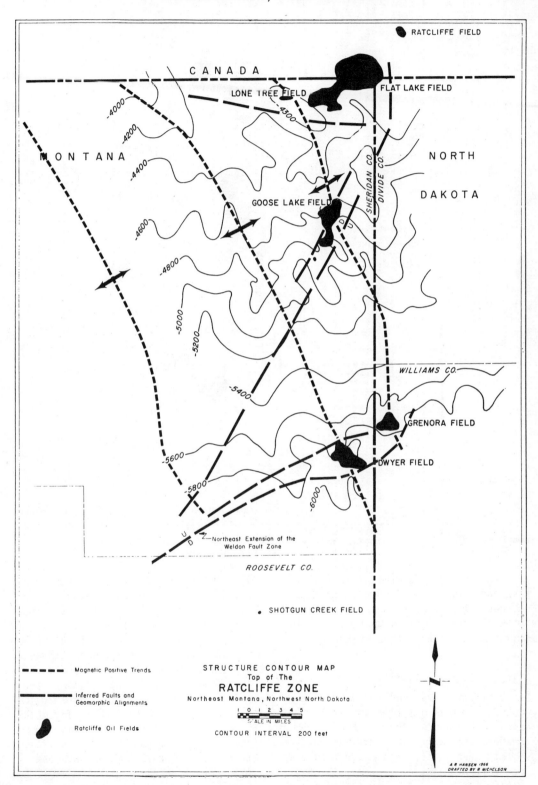

STRUCTURE CONTOUR MAP
Top of The
RATCLIFFE ZONE
Northeast Montana, Northwest North Dakota

CONTOUR INTERVAL 200 feet

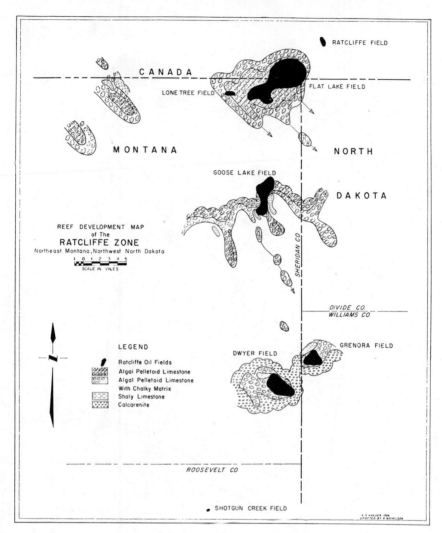

FIG. 4.—Reef-development map of Ratcliffe zone showing various lithologic types associated with the reef accumulations.

continued into the Oligocene (Howard, 1960, p. 17), and perhaps into the Pleistocene, because Howard (1960, Pl. 8) presents evidence that the ancestral Missouri River occupied a position with an orientation similar to that of the old fault pattern.

DWYER AND GRENORA FIELDS

If the Dwyer reef buildup is examined in detail, three distinct rock types are evident which are described and identified below. The spatial relations between these three rock types are illustrated in plan view by Figure 5 and in cross section by Figure 6.

Algal pelletoid reef limestone.—This lithologic type is the main reservoir rock and is by far the most abundant rock type in the overall reef complex throughout the area. Attention is called to the broken and fragmented nature of some of the individual limestone pellets shown in Figures 7 and 8. These shattered pieces commonly occur in distinct lenses, indicating the existence of periodic violent storm action that interrupted the normal development of these pellets. Most of the broken pellets have been enclosed completely by a subsequent algal layer, which reflects a resumption of the algal environmental conditions.

In a very few places, parts of cores cut in the

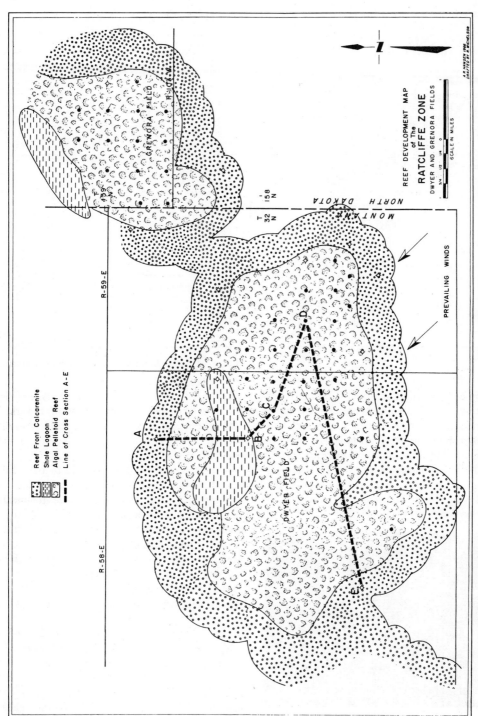

FIG. 5.—Reef-development map of Ratcliffe zone, Dwyer and Grenora fields, Montana and North Dakota, showing lithologic variations within reef development. Location of section A–E, Figure 6, is shown.

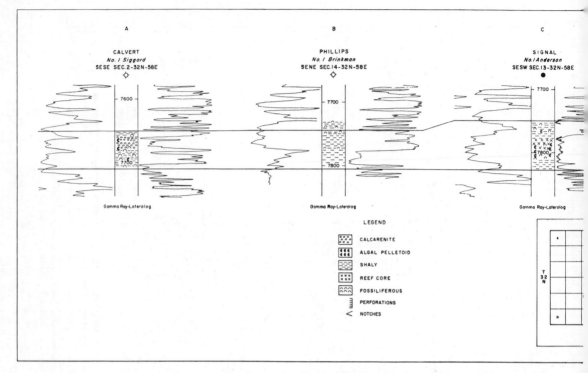

FIG. 6.—Cross section through Dwyer field, Montana, showing

Ratcliffe zone contain a massive, indistinctly bedded interval of porous limestone material that is designated reef core. Figure 9 is a typical photomicrograph of this lithologic type. Generally no positive fossil identifications can be made because the material has undergone varying degrees of reworking by burrowing organisms and subsequent alteration by recrystallization.

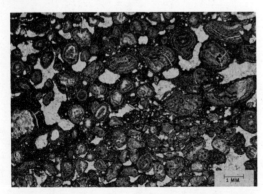

FIG. 7.—Algal pelletoid reef, Dwyer field, Montana. Note composite pellets, and broken pellets with subsequent overgrowths.

Reef-edge calcarenite.—This rock type (Fig. 10) is a heterogeneous assortment of reworked algal pellets, and argillaceous and shaly lagoonal material, including nearby open-marine fossil types and rock material.

In the Dwyer field, it is possible to trace this detrital high-energy deposit completely around the producing area. The coarsest-size fragments are predominantly along the southeast edge, which must have been absorbing the energy from the prevailing southeast winds. On the lee side of these reefs, the calcarenite is of similar composition, but the clastic components are smaller.

Of particular interest is a variety of endothyrid Foraminifera which has been observed only in the highly agitated calcarenite zone that surrounds the algal reef. This foraminifer developed a delicately whorled test that does not seem well suited to withstand the impact and abrasive energy that must have been so common in this part of the reef. Nevertheless, complete tests, and many nearly complete ones, are found throughout this detrital zone. It is suggested that these small forms may have been protected from destruction

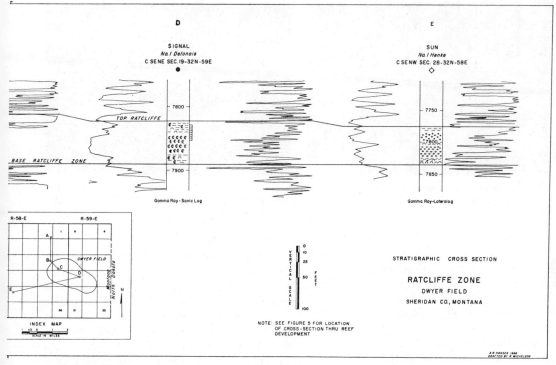

changes in lithologic character. Location of section shown on Figure 5.

by settling among fragments of much larger size.

Lagoonal shale deposits.—These quiet-water deposits consist of shaly and argillaceous limestone and are found commonly in the most protected area of the reef buildup. They do not represent a major constituent of the reef. Varying amounts of this rock type have been found in off-reef flank areas where restricted-energy environments are evident.

Figure 11 is a photomicrograph of a typical shaly lagoonal limestone mud deposit. A thin-section examination of this rock type indicates that virtually no identifiable fossils are present, except for a few brachiopod fragments and some irregularly interbedded carbonaceous material that may indicate the presence of former plant life.

STRUCTURE AND TECTONICS OF DWYER AND GRENORA FIELDS

Isopachous studies by the writer indicate that the present structural configuration of the Dwyer and Grenora producing fields is the result of reef growth and fault activity during Mississippian time. This growth undoubtedly was in-

creased or retarded periodically by intermittent activity along the fault-fracture system that passes through the two producing areas.

During the drilling of the Dwyer and Grenora fields, the writer was present at most of the wellsites to inspect and sample the cores. Although no vertical displacement can be demonstrated from mechanical-log correlations, slickensides, highly fragmented cores, and mineralization along

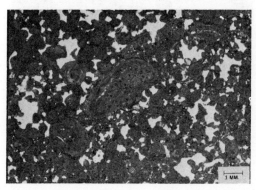

FIG. 8.—Algal pelletoid reef, Dwyer field, Montana. This is well D on Figures 5 and 6.

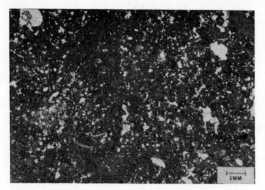

Fig. 9.—Reef core, Dwyer field, Montana. Note lack of identifiable fossils or algal pellets. Well C on Figures 5 and 6.

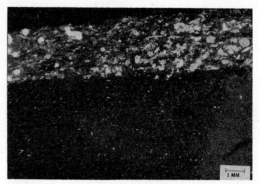

Fig. 11.—Shaly, calcareous mudstone, Dwyer field, Montana, showing lens of light-colored fossil fragmental material. Well B on Figures 5 and 6.

open vertical fractures indicate that post-depositional movement has occurred. Strong local dips and bedding-plane offsets as small as 1 in. (more or less) underlie undisturbed laminations and suggest displacement synchronous with deposition.

According to the writer's interpretation, these core data, coupled with an analysis of the overall regional tectonic framework, provide a reasonable explanation for the alignment of Ratcliffe productive trends shown on Figure 3.

FUTURE EXPLORATION

Within the area of Ratcliffe porosity development indicated on the index map (Fig. 1), additional oil discoveries are certain to be found. It is predicted that much of the future production from the Ratcliffe will be located parallel or on trend with current accumulations. Additional production also is predicted for the Devonian, Silurian, and Ordovician rocks whose depositional environments were very similar to that of the Mississippian.

CONCLUSIONS

The algal pelletoid reef development in the Ratcliffe zone was controlled primarily by small movements along older trends of weakness that were reactivated during Ratcliffe time. The relative direction of movement on some of these faults was reversed one or more times during the Paleozoic, thus masking this economically important tectonism from seismic and surface geologic exploration.

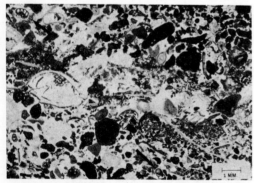

Fig. 10.—Calcarenite, Dwyer field, Montana, showing poorly sorted fossil fragments (brachiopods, crinoids, bryozoans, etc.), pellets, and limestone pellets. Light matrix is anhydrite and calcite. Well E on Figures 5 and 6.

REFERENCES CITED

Collier, A. J., and M. M. Knechtel, 1939, The coal resources of McCone County, Montana: U. S. Geol. Survey Bull. 905, 80 p.

Howard, A. D., 1960, Cenozoic history of northeastern Montana and northwestern North Dakota: U. S. Geol. Survey Prof. Paper 326, 107 p.

Norwood, E. E., 1965, Geologic history of central and south-central Montana: Am. Assoc. Petroleum Geologists Bull., v. 49, no. 11, p. 1824–1832.

THE AMERICAN ASSOCIATION OF PETROLEUM GEOLOGISTS BULLETIN
V. 51, NO. 10 (OCTOBER, 1967), P. 1959-1978, 27 FIGS.

CACHE FIELD—A PENNSYLVANIAN ALGAL RESERVOIR IN SOUTHWESTERN COLORADO[1]

R. S. GRAY[2]
Santa Barbara, California

ABSTRACT

Oil and gas accumulation in the Cache field in southwestern Colorado is chiefly in reservoirs developed in a northwest-trending group of "stacked" algal carbonate mounds and related carbonate facies in the Ismay zone (Des Moines age) of the Paradox Formation. Cache field is the most productive Pennsylvanian field in the Colorado part of the Paradox basin.

Characteristic cyclic deposition which is developed in the zones of the Paradox Formation is shown on a smaller scale within the Ismay of the Cache field. The Ismay has an average thickness of 180 ft and consists of nine carbonate lithofacies, anhydrite, siltstone, and black shale, which are recognized in one or more of the three cycles in the Ismay. The carbonate lithofacies are: (1) algal calcirudite breccia; (2) calcirudite breccia; (3) foraminiferal-pelletal limestone; (4) laminated boundstone; (5) crinoidal limestone; (6) "earthy" dolomite; (7) carbonate mudstone; (8) shelly mudstone; and (9) "evaporitic" dolomite.

The depositional environment of the normal-marine buildup cycle was dominated by the formation of algal mounds in relatively shallow water where phylloid *Ivanovia* algae grew in profusion on broad banks. In slightly deeper water flanking the mounds, and in channelways, fossiliferous carbonate mudstone was deposited. One facies (boundstone) probably formed in intertidal conditions.

The evaporitic cycle which followed the normal-marine buildup cycle probably was begun by the lowering of sea level. During the evaporitic cycle, "evaporitic" dolomite and massive anhydrite (gypsum) were deposited in an intertidal-supratidal environment.

Favorable zones of porosity and permeability are present within the algal calcirudite breccia and the dolomitized shelly mudstone. Porosity is predominantly secondary (early diagenetic) and is the result of leaching of the algae and solution brecciation (vugs) in the algal limestone, and leaching and recrystallization to dolomite in the dolomitized limestone.

INTRODUCTION

The Cache field in southwestern Colorado is the most productive field in the Colorado part of the Paradox basin. Exploration in the area for a northwest-southeast-trending lenticular Pennsylvanian algal carbonate-mound buildup of Des Moines age, similar to the Aneth and Ismay-Flodine Park algal-mound buildups, led to the drilling of the C. L. Veach No. 1 in the NW¼ NW¼, Sec. 2, T. 34 N., R. 20 W., Montezuma County, Colorado, in October 1964 (Fig. 1). On completion, this well flowed 1,434 bbl of oil per day from the Ismay zone of the Paradox Formation. Production is from porous strata in a group of

"stacked" biohermal carbonate mounds and local zones of leached fossiliferous and replacement dolomite. A total of 22 wells has been drilled since 1964, of which 18 have produced oil and 4 have been dry. The 45° API gravity Ismay oil is produced from an average depth of 5,620 ft on 40-acre spacing. Cumulative production to June 1, 1967, was 1,976,071 bbl of oil. Estimated recoverable reserves are 4.6 million bbl of oil with an original GOR of 918.

It was recognized early in the development of Cache field that the decline in the primary oil reserves would require that a secondary recovery program be initiated as soon as the field limits were outlined. A coring program was begun in order to gain a detailed knowledge of the different lithologic units and to define the lateral and vertical extents of the producing and nonproducing zones in the field. This program also increased the understanding of the sedimentary processes involved in development of the Pennsylvanian algal mounds.

GENERAL GEOLOGIC SETTING

Several papers have been written on the Pennsylvanian of the Paradox basin since the 1954 discovery of oil in the Desert Creek field in

[1] Revised from a manuscript read by K. E. Carter before the Rocky Mountain Section of the Association at Denver, Colorado, on October 27, 1966. Manuscript received and accepted, June 9, 1967.

[2] Assistant professor, Santa Barbara City College. The writer thanks Pan American Petroleum Corporation for permission to publish this paper. J. L. Severson, M. Smith, E. Strawn, D. F. Toomey, and F. Horacek of Pan American Petroleum Corporation, and P. Choquette of Marathon Oil Company, read the manuscript and offered valuable suggestions; their assistance is gratefully acknowledged. Special appreciation is due to K. E. Carter who helped the writer during the early stages of the paper. Drafting is by J. Jones.

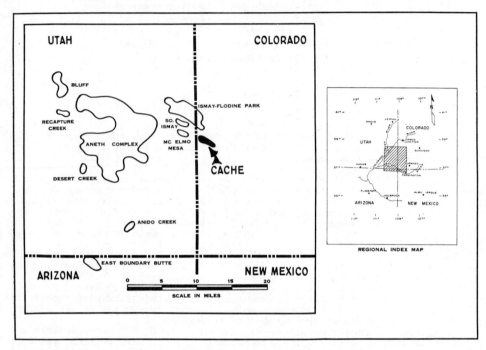

FIG. 1.—Index map of Pennsylvanian oil fields, Four Corners area.

southeast Utah (Fig. 1). Most of these not only summarize the general and regional stratigraphy and structural relations of the Pennsylvanian in the Paradox basin (Herman and Barkell, 1957; Peterson, 1959, 1966; Peterson and Ohlen, 1963), but also point out the importance of the algal carbonate mounds as the major carbonate reservoir facies in the Pennsylvanian. Recent papers by Choquette and Traut (1963), Elias (1963), and Pray and Wray (1963) present very ably the more local and detailed characteristics of the Pennsylvanian carbonate reservoirs and their relation to the enclosing facies. The reader is referred to these papers for a more detailed summary of the Pennsylvanian of the Paradox basin.

REGIONAL SETTING

Choquette and Traut (1963, p. 159) summarized the regional aspects of the Pennsylvanian of the Paradox basin as follows.

... they [Ismay carbonate buildups] lie along the northern margin of what appears to have been a broad, relatively shallow marine platform during Middle Pennsylvanian time. This platform, which Wengerd [and Matheny] (1958) called the "southwest shelf," formed a distal part of the Paradox basin, south and west of the axial part of the basin in which a thick section of evaporites was deposited during the Middle Pennsylvanian. Tectonically active areas bordered the shelf on the south (Defiance uplift), southwest (Kaibab uplift), and northwest (Emery-San Rafael uplifts). These areas, and the Uncompahgre land mass east of the Paradox basin, apparently shed debris intermittently into the shelf province, periodically interrupting the cyclic sedimentation (Herman and Barkell, 1957) of marine carbonates, evaporites, and other sediments. Carbonate deposition prevailed over much of the southwest shelf during the Middle Pennsylvanian and most known production from the Paradox Formation comes from carbonate rocks of this province.

The Pennsylvanian of the Paradox basin consists of the basal Molas Formation, which is a reworked red soil regolith of the Mississippian, and the Hermosa Group (Baars et al., 1967). The Hermosa is divided into the Pinkerton Trail Formation (lower Hermosa), Paradox Formation, and the Honaker Trail Formation (upper Hermosa). The Paradox Formation contains abruptly changing lateral facies that thicken considerably basinward. Vertically, these facies are repeated and have been considered to be cyclic in nature (Elias, 1963; Peterson and Ohlen, 1963). On the basis of this cyclic nature, the Paradox Formation has been subdivided, from oldest to youngest, into "pay zones"—Barker Creek zone,

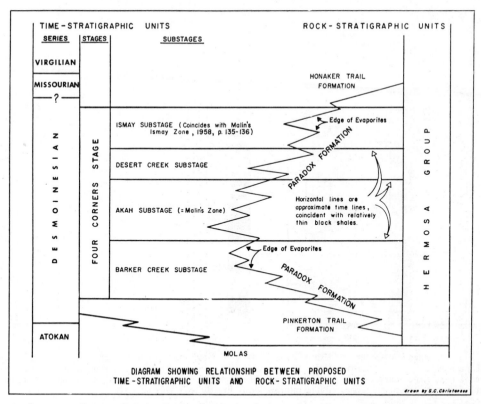

FIG. 2.—Summary diagram, Paradox basin, showing revised stratigraphic nomenclature proposed by Baars *et al.* (1967, p. 401). Permission granted by Baars *et al.* to use diagram directly from their paper.

Akah zone, Desert Creek zone, and Ismay zone[3] (Fig. 2). These zones can be correlated on radioactivity logs across considerable distances because of the presence of widespread thin dark shale and dark carbonate mudstone beds which divide the Paradox interval into recognizable carbonate-anhydrite units. These fine-grained rocks appear to be time-stratigraphic units (Baars *et al.*, 1967). Several geologists (Elias, 1963; Peterson, 1966) have suggested that the orderly repetition of the cyclic layers may be the result of sea-level fluctuations during the Pennsylvanian.

LOCAL SETTING

In the Cache field the Ismay zone averages 180

[3] The reader is referred to the recent paper on the "Revised Stratigraphic Nomenclature of Pennsylvanian System, Paradox Basin," by D. L. Baars, J. W. Parker, and J. Chronic in the March 1967 issue of the AAPG *Bulletin* in which the authors suggest that the "pay zones" of the Paradox Formation should properly be called "substages." However, the word "zone" will be used in this paper.

ft in thickness, and has been subdivided by most geologists into lower and upper Ismay subzones. Each subzone is characterized by a basal black shale or a dark carbonate mudstone similar to the dark shale and carbonate mudstone beds that subdivide the Paradox Formation (Fig. 3). These shales are also considered to be time-stratigraphic units. Within the upper Ismay, a further subdivision is noticeable between the carbonate mounds and the flanking and overlying anhydrites; some geologists consider this to be a third subzone (Fig. 3).

The Cache field is an example of a Paradox field that shows the trapping to be primarily stratigraphic, with only minor structural control. The structural pattern in the Cache field prior to deposition of the Ismay carbonate unit, using as contour datum the top of the shale marker at the base of the Ismay carbonate zone, is shown in Figure 4. This figure suggests that the area was essentially flat. However, the present structural configuration of any Pennsylvanian datum proba-

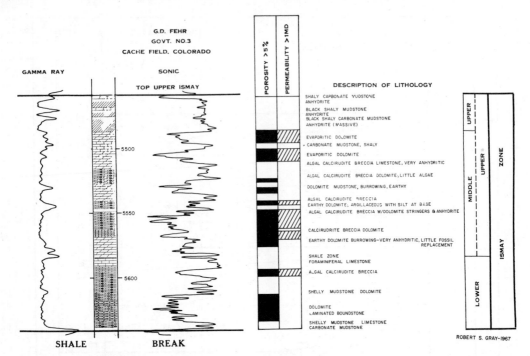

FIG. 3.—Schematic section of Texaco No. 3 G. D. Fehr Government well in Cache field showing lithologic and mechanical-log characteristics of Ismay zone from shale marker at base of carbonate lithologic types to base of next shale marker (top of Ismay zone). See Figure 7, cross section A-A', for visual description of lithologic types. Porosity and permeability information is directly from core analysis.

bly is not truly representative of paleostructure. The structural pattern, using as datum the top of the Ismay zone in the Cache field, shows the presence of about 50 ft of closure, as well as a "high" area which appears to be related directly to the development of the algal mounds (see Figs. 5 and 6). Figure 5, using as datum the top of the middle Ismay, shows even greater closure.

CACHE FIELD COMPLEX

Present control indicates that the Ismay zone in the Cache field is composed of several related lithologic facies, and that the lateral and vertical extents of these facies form a northwest-southeast-trending lenticular mound. Although the Paradox mounds, the brecciated carbonate units, and the vertical buildups are suggestive of a reef complex, most geologists (Elias, 1963; Pray and Wray, 1963; Choquette and Traut, 1963; Klement, 1967) have stated that the paleontological content (phylloid algae, crinoids, brachiopods, and foraminifers) of these mounds is indicative of bank deposits. Such organisms cannot act as

framework builders and, therefore, form structures in which the sediment massing is caused by biogenic baffling and a binding of the sediments. Many geologists believe a reef is built by the in-place growth of organisms which ecologically act as framework builders. The writer agrees that these structures are suggestive of bank deposits and not reefs.

A third term used commonly in discussion of Paradox carbonate units is "buildup." Merriam (1962, p. 73) recommends the use of this term instead of reef or bank because it has no genetic implication but can be used to refer to any extra, stray, or "super" carbonate (limestone) bed or beds in addition to the "normal" sequence as exemplified in rhythmic (cyclic) deposits.

LITHOFACIES CYCLES IN ISMAY ZONE

Choquette and Traut (1963, p. 169) noted that the lithologic units in the Ismay zone tend to be cyclically repeated in any particular vertical section. They delineated at least three cycles, namely (1) a normal-marine cycle containing a carbonate buildup (in this paper called the normal-ma-rine buildup cycle); (2) an evaporitic cycle; and (3) a dominantly normal-marine cycle found in sections between buildups along the productive trend. Locally, these cycles also succeed one another laterally within distances of only a few hundred feet.

In the Cache field, although parts of all three cycles are represented, the normal-marine buildup cycle is the dominant cycle in most of the wells (Fig. 7). In the edge wells, especially the Pan American No. G-2 Ute, No. J-2 Ute, and No. J-3 Ute, and the Vaughey and Vaughey No. 1 Vaughey-Veach, a transitional cycle is present in the lower Ismay between the normal-marine buildup

cycle and the normal-marine cycle between buildups, with lithologic units of both cycles overlapping and interfingering (Figs. 8, 9, 10). Another variation is present in the upper Ismay where lithologic units of the normal-marine cycle between buildups were deposited in a northeast-southwest-trending channelway between buildups (Fig. 7). The evaporitic cycle, which contains thick, massive, nodular anhydrite, is present in all the wells, but is thickest in wells flanking the normal-marine buildup.

Lithologic units of the Ismay zone in the Cache field are similar to lithologic types described by other authors in most of the Pennsylvanian fields of the Paradox basin. The terminol-

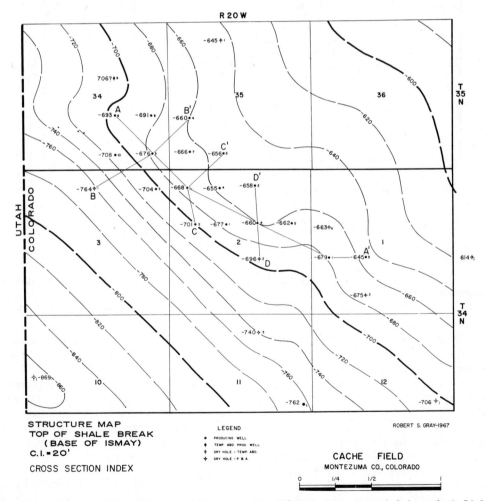

FIG. 4.—Structural map. Datum is base of Ismay carbonate lithologic types (top of shale marker). Little or no structure evident. Index to four stratigraphic cross sections is superimposed on structural map.

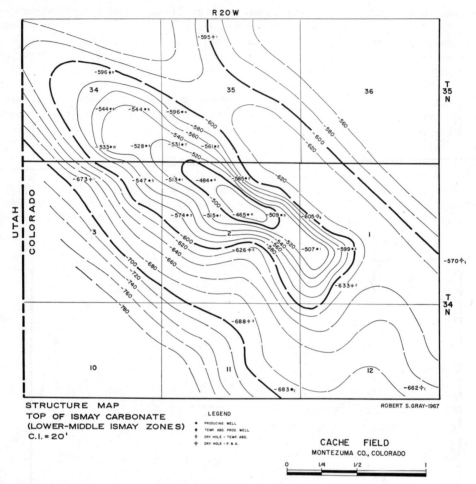

FIG. 5.—Structural map. Datum is top of Ismay carbonate (lower-middle Ismay) zone.

ogy and classification that are generally used for lithologic units of the Ismay zone have been described in the detailed study of the Ismay zone from the Ismay-Flodine Park field, Utah and Colorado, by Choquette and Traut (1963). The writer uses their terminology in this paper.

Within the Ismay zone, there are distinct lithologic and biotic variations that naturally divide the zone laterally into several lithofacies.

At least nine different carbonate lithofacies plus anhydrite, black shale, and siltstone have been recognized from detailed core studies. The nine carbonate lithofacies, named according to their distinctive component, are: (1) algal calcirudite breccia; (2) calcirudite breccia; (3) foraminiferal-pelletal limestone; (4) laminated boundstone; (5) crinoidal limestone; (6) "earthy"

dolomite; (7) dark carbonate mudstone; (8) shelly mudstone; and (9) "evaporitic" (earthy) dolomite (Fig. 3). Lithofacies units 1 through 5 are restricted to the Cache field normal-marine buildup cycle. Lithofacies 6 through 8 may be present within the normal-marine carbonate cycle throughout the area, even some distance from Cache field. Lithofacies 9 and anhydrite are related to the evaporitic cycle. Diagenetic changes, especially secondary dolomitization and anhydritization, cause numerous variations within the lithofacies. A discussion of the carbonate and anhydrite lithofacies follows.

Lithofacies of Normal-Marine Buildup Cycle

1. *Algal calcirudite breccia.*—This lithofacies, the main productive reservoir rock in the Cache

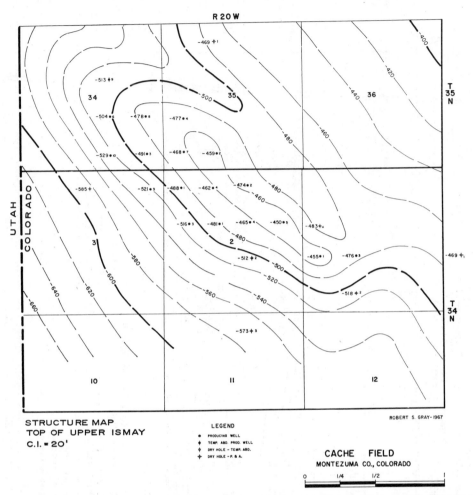

R 20 W

STRUCTURE MAP
TOP OF UPPER ISMAY
C.I. = 20'

LEGEND
● PRODUCING WELL
◆ TEMP. ABD. PROD. WELL
◇ DRY HOLE – TEMP. ABD.
✛ DRY HOLE – P. & A.

ROBERT S. GRAY- 1967

CACHE   FIELD
MONTEZUMA CO., COLORADO

0   1/4   1/2   1

FIG. 6.—Structural map. Datum is top of upper Ismay. At least 50 ft of closure is indicated. This thickness appears to be related directly to algal-mound buildup during Ismay time.

field, is characterized by the abundance of codiacean(?) phylloid (Pray and Wray, 1963, p. 209) algal material, primarily *Ivanovia* (Khvorova, 1946) plates occurring with micritic intraclasts (Fig. 11). It is a grain-supported, algal-plate boundstone[4] that consists of 45–55 percent micritic intraclasts, 15 percent sparry matrix, 5–7 percent pellets, and 15–22 percent biotic material. The most important biotic element is the phylloid alga *Ivanovia,* but foraminifers, brachiopods, and encrusting bryozoans also are abundant. Crinoidal debris, pelecypods, rugose corals, and questionable blue-green algae are scarce. General-

[4] The writer uses Dunham's (1962) classification for grain-supported and mud-supported carbonate rocks.

ly, the micritic intraclasts are dark gray, subangular to angular, breccia blocks which have been slightly rotated (Fig. 12). The intraclasts appear in thin section to be very fine, sparsely fossiliferous, crystalline masses (Fig. 13). Most of the fossils in the intraclasts are foraminifers. Typically, the phylloid algae acted as a baffling agent for the entrapment of carbonate mud, and some show *in situ* compaction brecciation.

The calcareous alga *Ivanovia* is rarely in an upright growth position but generally occurs as a jumbled mass of loosely packed plates, some of which have been badly fragmented. Many have been leached, recrystallized (Fig. 14), dolomitized, or replaced by anhydrite.

Choquette and Traut (1963) have suggested

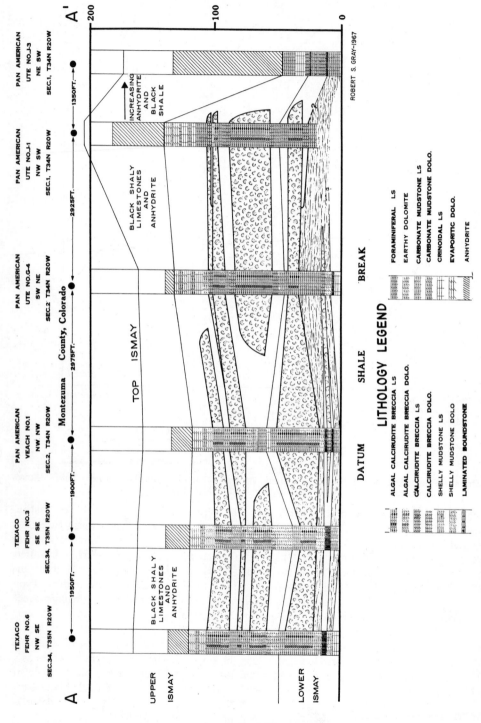

Fig. 7.—NW-SE stratigraphic cross section A-A'. The cross section strikes along trend of algal mounds (Fig. 4). "Stacked" biohermal mounds are delineated by bush pattern. The Pan American No. J-3 Ute illustrates the abrupt facies change to dolomitic carbonate mudstone within 40 acres. All wells except Texaco No. 6 G. D. Fehr Government have been cored. Vertical exaggeration is great. Graphic lithologic symbols apply to all stratigraphic cross sections in Figures 3, 7, 8, 9, and 10. Symbols not shown graphically are: (1) bush pattern–biohermal mounds; and (2) shaly (bark) pattern of lower Ismay.

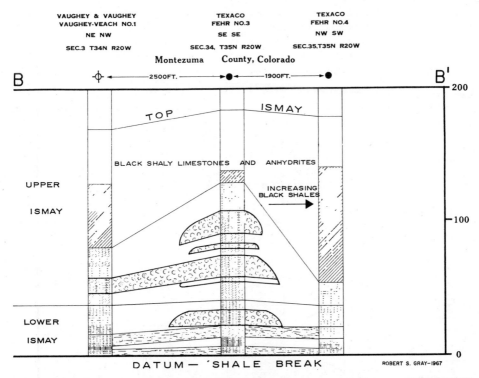

FIG. 8.—SW-NE stratigraphic cross section B-B'. Cross section cuts across linear trend of field (Fig. 4). Algal mounds are restricted to center well. Edge wells (Vaughey and Vaughey No. 1 Veach, and Texaco No. 4 G. D. Fehr Government) contain dark fossiliferous mudstone equivalent in part to algal mounds of center well. The shelly mudstone and related laminated boundstone are continuous in all wells. Massive nodular anhydrite is thicker in edge wells. Bark (shaly) pattern represents shelly mudstone facies. See Figure 7 for graphic lithologic legend.

that the phylloid algae grew in abundance on shallow banks in a warm shallow sea bordering the deeper Paradox basin. Sedimentation rates probably were higher on the shallow mounds than on the flanks, and the algal mounds built upward more rapidly because of the abundance of phylloid algae. Some of these algal mounds were built at different times and in different places on the shallow bank during the Ismay cycle (Fig. 7).

It is probable that in growth position the algae were a few inches high (Pray and Wray, 1963, p. 216) and consisted of a rather limited number of relatively broad, somewhat rigid and brittle "leaves" or fronds suggesting a loosely packed head of lettuce. It is believed that the "leaves" could not withstand a high degree of current and wave turbulence and that they thrived in a zone beyond or away from wave action (D. F. Toomey, personal commun.). These algae probably could not have provided the framework or encrustation necessary to support a growing reef.

However, they were probably very efficient sediment producers, as shown by their ability to encrust (calcify) their "leaves" and to provide an effective sediment-baffling mechanism. In this sense, these algae probably are analogous with the modern-day *Halimeda* and the sediment-baffling *Thalassia* (turtle-grass). Klement (1966, p. 367) has indicated that *Ivanovia* probably was very efficient in trapping suspended carbonate sediments. The fine carbonate mud and silt that were trapped by them were added to the mud substratum and helped to build the mound.

Brecciation of the mud to form micritic intraclasts may have been controlled by the compaction and leaching of the algal "leaves" and possibly by subaerial desiccation. Skeletal debris within this facies generally shows no wear or abrasion. Many of the fossils are unbroken, indicating that movement of the sediments was minor. The buildup usually was "killed off" or drowned when sedimentation could not keep up with subsidence

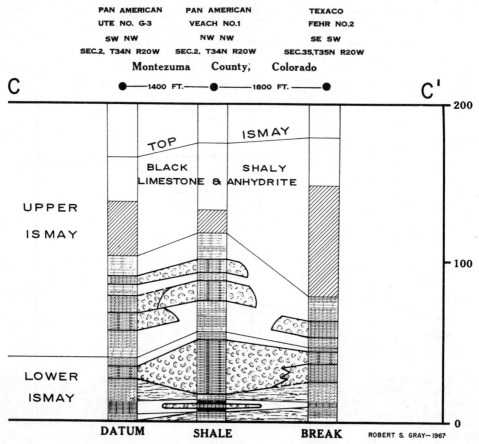

FIG. 9.—SW-NE stratigraphic cross section C-C'. Similar to cross section B-B' but crosses field at mid-point (Fig. 4). Pan American No. 1 Veach has thickest algal mound in lower Ismay. This well has dark carbonate mudstone equivalent to algal mound buildup in upper Ismay in Texaco No. 2 G. D. Fehr and in Pan American No. G-3 Ute. In upper Ismay in Pan American No. G-3 Ute, the calcirudite breccia grades laterally into algal calcirudite breccia in Pan American No. 1 Veach. Massive nodular anhydrite is thicker in edge wells. See Figure 7 for graphic lithologic legend.

or when the sea level dropped rapidly, exposing the mound. Where the buildup was drowned, encrusting foraminifers and micritic mud slowly engulfed it.

Secondary dolomitization has affected some of the mounds, especially near the edge of Cache field. The edge wells usually have thinner buildup zones and have been thoroughly dolomitized. However, secondary anhydrite has effectively sealed most of the porosity in these wells. A few of the algal calcirudite breccia zones in the middle of the field are dolomitized and have excellent porosity, as in the Texaco No. 3 G. D. Fehr Government well (Fig. 3).

2. *Calcirudite breccia.*—This lithofacies is

similar in all aspects to the algal calcirudite breccia and is probably a subfacies of it. These rocks are characterized by an increased amount of micritic intraclasts and a scarcity of phylloid algae. In fact, the micritic intraclasts make up 65–75 percent of the rock with 7–10 percent skeletal material, 3–5 percent pellet grains, and 15 percent matrix material (usually carbonate mud and anhydrite; Fig. 15). The biota is also similar to that of the algal calcirudite breccia. Vugs and void spaces are rare; much of the phylloid algae has been recrystallized to sparry calcite leaving little or no porosity. Generally, this unit grades laterally and vertically into the algal calcirudite breccia, but in some wells there is no associated

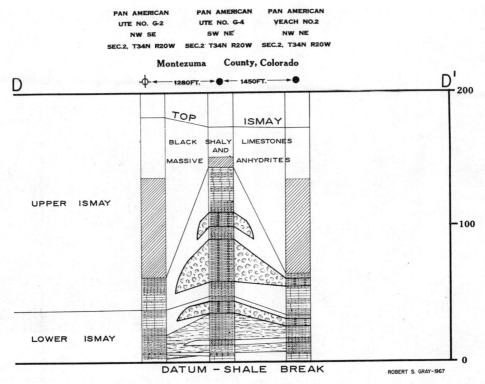

PAN AMERICAN   PAN AMERICAN   PAN AMERICAN
UTE NO. G-2   UTE NO. G-4   YEACH NO.2
NW SE   SW NE   NW NE
SEC.2, T34N R20W   SEC.2 T34N R20W   SEC.2, T34N R20W

Montezuma   County, Colorado

FIG. 10.—SW-NE stratigraphic cross section D-D′. Similar to cross section B-B′ but crosses southeastern part of field (Fig. 4). Pan American No. G-2 Ute penetrated dark carbonate mudstone equivalent to algal-mound buildup in lower Ismay. Shelly mudstone is thicker eastward whereas algal-mound buildup in lower Ismay is thinner (Pan American No. G-4 Ute). Massive anhydrite is much thicker in edge wells. See Figure 7 for graphic lithologic legend.

algal calcirudite breccia. Calcirudite breccias are more common in the edge wells, suggesting that the phylloid algae grew in less abundance on the flanks of the buildups. The presence of a large amount of secondary anhydrite in this unit further reduces the porosity.

3. *Foraminiferal-pelletal limestone.*—Foraminiferal-pelletal limestone is characterized by the abundance of pellets of structureless cryptocrystalline calcite and abundant foraminiferal skeletal grains generally set in a carbonate-mud matrix. In thin section, the facies ranges from a foraminiferal grain-supported packstone (Fig. 16) to a fusulinid wackestone (Fig. 17).

The foraminifers are plentiful (50 percent or more of the rock) and diversified. Encrusting types such as *Tuberitina*, *Tetrataxis* (Fig. 14), and *Hedraites*, and calcareous mobile types such as *Climacammina*, *Bradyina*, *Globivalvulina*, *Endothyra*, and fusulinids (*Fusulina* sp.) are abun-

dant. Agglutinated mobile and encrusting types such as *Hyperammina* and *Minammodytes*(?) also are present. No attempt is made to identify the foraminifers to the species level.

Although the foraminifers occur throughout most of the other facies, they are so prolific in this facies as to exclude most other skeletal debris. In a few places, parts of this facies are dolomitized and have excellent porosity. Secondary anhydrite is scarce.

4. *Laminated boundstone.*—This lithofacies is nearly everywhere dolomite and is recognized by the presence of laminated breccia blocks that appear to represent *in situ* mud-cracked slabs and transported chips of carbonate mudstone associated with algal mat laminations and burrows (Fig. 18). The blocks usually are broken, offset, and the fractures infilled by skeletal debris (brachiopods and pelecypods) and pellets in a silt- and clay-size dolomite matrix, suggesting formation in a tidal-

FIG. 11.—Algal calcirudite breccia. Gray, micritic intraclasts in matrix of fossil debris and sparry cement. Most fossil debris is *Ivanovia* fragments which acted as sediment traps for micritic intraclasts. Excellent porosity caused by leached *Ivanovia* fragments. Formation of vugs and voids is due in part to collapse breccia. Polished core, Pan American No. J-1 Ute well, depth 5,588 ft.

FIG. 12.—Algal calcirudite breccia. Slightly dolomitized, gray, micritic intraclasts are supported by tan matrix of *Ivanovia* fragments which have been dolomitized. Leached fragments give excellent porosity. Interparticle porosity is caused by secondary dolomitization of sparry calcite. Rock fabric developed despite dolomitization. Polished core, Texaco No. 3 G. D. Fehr Government well, depth 5,598 ft.

flat environment where the flat was wet and dried intermittently. Dolomitization is secondary and probably occurred shortly after formation of the carbonate mud. Porosity is usually poor because of secondary anhydrite infill and replacement of the fossils.

The laminated boundstone is common in the lower Ismay and is everywhere associated with the shelly mudstone lithofacies. It is probably a subfacies of the shelly mudstone which has been subaerially exposed in the intertidal zone. It is found locally in the upper Ismay in association with some calcirudite breccias.

5. *Crinoidal limestone.*—Lithologically, this facies contains abundant skeletal grains of crinoidal columnal debris in a dense, black carbonate-mudstone matrix. In thin section, the crinoidal columnals appear to be randomly oriented in the mud matrix to form a crinoidal wackestone. In some

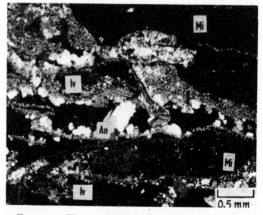

FIG. 13.—Thin-section photomicrograph of grain-supported, algal calcirudite breccia packstone of *Ivanovia* fragments (Iv). Fragments are leached or partly recrystallized to sparry calcite. Anhydrite crystals (An) are common and in some wells have effectively sealed porosity. Dense carbonate intraclasts (Mi) are dolomitized or fossiliferous in very few places. Black space is porosity. Pan American No. J-1 Ute well, crossed nicols, depth 5,590 ft.

wells the crinoidal debris is so densely packed that it forms a crinoidal packstone (Fig. 19). The crinoidal columnals range from silt-size fragments to fragments ¾ in. in diameter, but average ¼ in. in diameter. Although other fossil material usually is scarce, brachiopods, fusulinids, and other foraminifers are found in abundance locally.

Crinoidal limestone probably formed mostly in quiet water below wave base. The crinoidal material is abraded in very few places, a fact which implies that it was not transported far. The dark carbonate mudstone matrix (Fig. 20) in close proximity to the lighter algal breccia carbonate suggests that the crinoidal limestone was deposited in deeper and more stagnant water. This facies generally is found along the flanks of the mounds and usually is associated with the dark dense carbonate mudstone.

The limestone rarely is dolomitized. Sparry calcite has replaced most of the crinoidal columnals, but, locally, secondary anhydrite has effectively replaced all the crinoidal debris and has partly replaced the carbonate–mud matrix. The rock is dense and nonporous.

Lithofacies Not Restricted to Normal-Marine Buildup Cycle

6. *"Earthy" dolomite.*—Tan, saccharoidal, argillaceous, silty, dolomitic mud is scattered

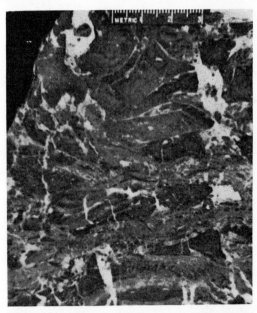

FIG. 15.—Calcirudite breccia. Gray, micritic intraclasts dominate (60–80%) this rock. Matrix material is usually micritic limestone which may contain numerous fossil fragments. *Ivanovia* fragments are rare. Secondary fractures are filled with anhydrite. This rock type may grade vertically and horizontally into algal calcirudite breccia. Porosity is poor as there is usually no leaching. Pan American No. G-3 Ute well, polished core, depth 5,533 ft.

throughout the normal-marine buildup cycle but is not restricted to this cycle. The fine crystalline character of the silt-size dolomite rhombs gives

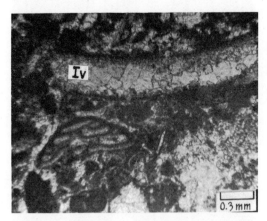

FIG. 14.—Thin-section photomicrograph of mixed-fossil-intraclast (algal calcirudite breccia) wackestone. Section is from edge well in which *Ivanovia* fragments (Iv) have been rotated, rounded, and recrystallized to sparry calcite. Fossil fragments of bryozoans, foraminifers (*Tetrataxis* (?)), crinoids, and micritic intraclasts are common. They show partial reworking. No porosity. Pan American No. 2 Veach, crossed nicols, depth 5,577 ft.

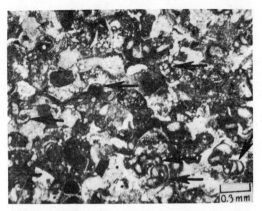

FIG. 16.—Foraminiferal-pelletal packstone. Thin-section photomicrograph shows grain-supported limestone with high (40–50%) encrusting and mobile foraminifers (note arrows). Fine opaque specks are pelletal material. Matrix is micrite and sparry calcite. No porosity. Pan American No. 2 Veach, plane-polarized, depth 5,574 ft.

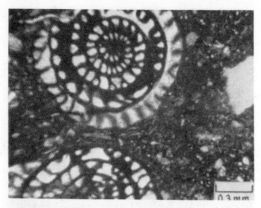

FIG. 17.—Fusulinid wackestone. Thin-section photomicrograph of dark carbonate mudstone where grains are mostly fusulinids. Debris of crinoids and bryozoans are accessory. Pan American No. G-2 Ute, crossed nicols, depth 5,720 ft.

FIG. 19.—Crinoidal packstone. Thin-section photomicrograph of crinoidal limestone in which close packing has excluded mudstone matrix. Local zones in the crinoidal carbonate mudstone may be considered grainstone. Pan American No. 1 Veach, crossed nicols, depth 5,510 ft.

the dolomite an "earthy" appearance. As seen in thin section, the rock is grain-supported, and is composed of homogeneous, silt-size dolomite

rhombs. The most characteristic sedimentary feature is the thin, cross-laminated stratification and flow-banding, strongly suggesting the formation of "penecontemporaneous" dolomite in shallow agitated water where the sediments occasionally

FIG. 18.—Laminated boundstone. Dolomitized fragmented laminae or micritic chips usually are broken and offset or slightly rotated. The fractures commonly are filled with silt-size dolomite grains, dolomitized fossil debris, or pelletal material. The offset appearance suggests formation and desiccation on a tidal flat, and is associated with shelly mudstone in lower Ismay. Anhydrite has filled vugs and voids, thus causing decrease in effective porosity. Texaco No. 3 G. D. Fehr Government well, polished core, depth 5,631 ft.

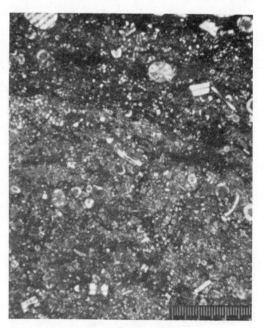

FIG. 20.—Crinoidal limestone. Common facies of dark carbonate mudstone which in some places is calcarenite of crinoidal debris. Most white specks are fragments of crinoids and in many wells have been recrystallized to sparite or replaced by anhydrite. The micritic groundmass rarely is dolomitized. Pan American No. 2 Veach, polished core, depth 5,554 ft.

were stirred up (Fig. 21). Fossils are conspicuously absent, but lens-like features in the dolomite are common locally and may indicate burrowing activity. Nodular anhydrite blebs interrupt the homogeneous appearance. "Earthy" dolomite within the normal-marine buildup cycle grades vertically into dolomitized fossiliferous limestone. Intercrystalline porosity is high but permeability is low.

7. *Carbonate mudstone.*—Dark-gray or brown to black carbonate mudstone, which is almost entirely limestone, is dense and fetid and resembles black shale (Fig. 22). In most places it contains black shale and siltstone stringers, which range from $\frac{1}{4}$ to 6 in. in thickness. In thin section, the carbonate grains are commonly 7 to 10 microns but may be as much as 40 microns in size. The carbonate mudstone contains a high percentage of disseminated organic matter and sulfides, mostly pyrite. Near clastic shale, the carbonate mudstone tends to be cross-laminated and contains clastic clay and silt-size quartz grains.

Fossils common in the mudstone are fusulinids, crinoids, bryozoans, thin-shelled brachiopods, and

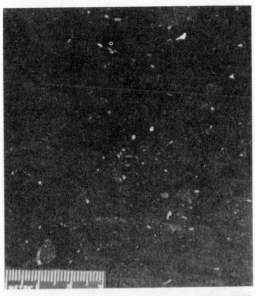

Fig. 22.—Dark carbonate mudstone. Dense, fetid, organic, slightly fossiliferous rock. Grades into fossiliferous carbonate mudstone. Thin, black shale streaks and a few thicker shale stringers are interbedded with carbonate mudstone. Quartz silt grains and pyrite laminae are scarce. Rock is dolomitized in very few places and is effective in preventing vertical movement of fluids. Texaco No. 3 G. D. Fehr Government, polished core, depth 5,473 ft.

some conodonts, sponge spicules, and ostracods. This mudstone is almost wackestone, texturally, and may grade vertically into fossiliferous mudstone, specifically foraminiferal limestone and crinoidal limestone.

This facies developed in a quiet-water environment in which the water was sufficiently oxygenated to support calcitic forms of life, but was of sufficiently low energy to cause mildly reducing conditions. These mudstone units are common in the lower Ismay throughout southwestern Colorado.

8. *Shelly mudstone.*—The shelly mudstone is a light-colored, mud-supported, moderately fossiliferous limestone in which the fossils consist almost entirely of pelecypods and brachiopods (Fig. 23). Texturally, it is a fossiliferous wackestone (Fig. 24). The shelly mudstone appears mainly near the base of the lower Ismay above a carbonate mudstone unit, and is extensive throughout southwestern Colorado. The shelly mudstone and a laminated boundstone facies associated with it, which probably formed in supratidal areas of low relief, may have been the base

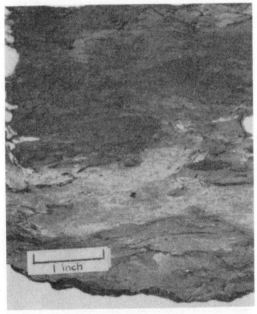

Fig. 21.—"Earthy" dolomite. Tan, argillaceous, finely crystalline dolomite has "earthy" appearance. Commonly it has thin laminae and cross-laminae which may contain organic wisps. Nodular anhydrite blebs are abundant. No fossil material. Porosity is excellent; permeability is poor. Texaco No. 2 G. D. Fehr Government, polished core, depth 5,558 ft.

FIG. 23.—Shelly mudstone. Light-colored, moderately fossiliferous, mud-supported rock. Most fossils are spiriferoid and productoid brachiopods or pelecypods. Crinoids and bryozoans are rare. This facies is common in lower Ismay in southwestern Colorado and may be substratum that started algal mounds. Pan American No. 2 Veach, polished core, depth 5,583 ft.

extent almost everywhere. Where the dolomitization is complete, the fossils have been obliterated except where secondary anhydrite has replaced them. Most of the shells are well preserved; some are essentially unbroken, suggesting the formation of a fossiliferous carbonate mud in shallow, oxygenated, quiet water. Spiriferoid and productoid brachiopods and pectinoid pelecypods are the usual fossils, with crinoids, echinoids, bryozoans, and fusulinids common locally.

This facies is second only to the algal-mound buildup facies as a major productive zone. Porosity values are high and permeability values are good due to the dolomitization of the mudstone and the selective leaching of the fossils (Fig. 25).

Although the rock types are diverse, exhibiting variations in mineral composition, in type of skeletal and nonskeletal grains, in textures, and in structures, the depositional environment is interpreted to have been normal marine. The lithofacies represent sediments deposited in an environment ranging from slightly stagnant to well-oxygenated, quiet water in a shallow sea and from normal to somewhat higher than normal salinity. Such an environment is thought to be similar to that of the modern-day Florida Bay-Bahama Banks area.

Lithofacies of Evaporitic Cycle

9. *"Evaporitic" dolomite.*—"Evaporitic" dolomite is similar in all respects to the "earthy"

on which the algal mounds formed during transgression.

The shelly mudstone is dolomitized to some

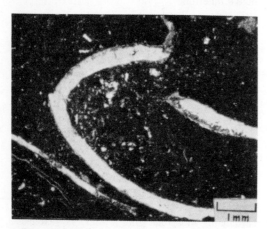

FIG. 24.—Shelly wackestone. Thin-section photomicrograph of shelly mudstone showing partially articulated brachiopod in micritic carbonate groundmass. Shelly mudstone has been dolomitized. Pan American No. G-2 Ute, plane polarized, depth 5,724 ft.

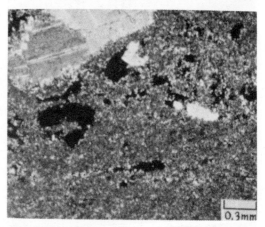

FIG. 25.—Shelly dolomite. Thin-section photomicrograph of shelly mudstone by dolomite. Most fossils have been leached. Anhydrite crystals have replaced some fossils and dolomite. Dolomite crystals are silt-size. Excellent interparticle and vuggy porosity and permeability. Pan American No. 1 Veach, crossed nicols, depth 5,564 ft.

dolomite already described, but it appears to be associated with the evaporitic cycle. At the base of all massive nodular anhydrite units (Figs. 8, 9), a tan, saccharoidal, very silty and shaly, anhydritic dolomite is present. It is homogeneous in grain size (silt-size to very fine sand-size) and commonly shows little or no cross-lamination or bedding. Organic wisps and nodular anhydrite blebs are disseminated throughout the dolomite.

In wells containing the normal-marine buildup cycle as shown in Figure 3, two "evaporitic" dolomite units are found near the top of the upper Ismay carbonate zone and seem to indicate a change from the normal-marine buildup cycle to the evaporitic cycle. In some wells, these dolomites are productive zones. Intercrystalline porosity values are high, but the corresponding permeability values generally are low.

10. *Anhydrite.*—Anhydrite is the most important noncarbonate mineral. It is commonly associated with the "earthy" and "evaporitic" dolomites and nodular anhydrites, but also is abundant as an accessory mineral in the other carbonate facies. Three types of anhydrite are recognized (Murray, 1964, p. 515)—(1) nodular, (2) void-filling, and (3) replacement. Commonly, the void-filling anhydrite grows within vugs and solution cavities in the algal calcirudite breccia, calcirudite breccia, and laminated boundstone, and in interparticle space in all the carbonate facies. Anhydrite commonly replaces bioclastic material. Some fossils were recrystallized, then altered to dolomite, and finally replaced by anhydrite. Relict crystals of dolomite are present in the anhydrite. Uncommonly, anhydrite replaces the entire rock.

The most important type of anhydrite is the nodular type. Although it is found throughout the Ismay zone as blebs in other lithofacies, it is thickest in the massive anhydrite section in wells flanking the northwest-trending mounds (Figs. 8, 9, 10).

The nodules, which are white to light gray, range widely in diameter from less than ¼ in. to 5 in. They are varied in shape although spherical shapes are most common. Many occur as individual blebs "floating" in a matrix of wavy and laminated dolomitic mudstone or organic black shale. As the nodules increase in quantity, they form a complex mass of closely packed nodules separated only by thin wisps of dolomite or organic residue (Fig. 26). These nodules appear to be somewhat

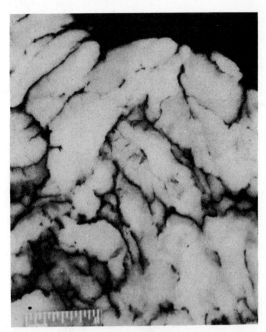

FIG. 26.—Massive nodular anhydrite. Closely packed white anhydrite nodules separated by wisps of dolomite and organic residue. Individual nodules composed of anhydrite crystals. Pan American No. 2 Veach, polished core, depth 5,472 ft.

bedded, partly due to grooving caused by the core barrel, and have been mistakenly called bedded anhydrite. During formation, the nodules evidently grew at the expense of the enclosing sediments. The enclosing dolomite or shale shows compaction features as well as a "wrapping" effect around the anhydrite nodules (Fig. 27).

The nodules, as seen in thin sections, consist of a felty mass of finely crystalline anhydrite laths. The laths are elongate parallel with their *c* axis and are distally spear-shaped. Near the contact with the enclosing sediments, many laths tend to be aligned parallel with the curved margin of the nodule. Most of the nodules are free of foreign material.

Environmental considerations of evaporitic cycle.—No conclusive stratigraphic proof has been found as to the relative time of deposition of the massive nodular anhydrite in the flank wells and the "stratigraphically equivalent" carbonate lithofacies of the normal-marine buildup cycle, but it appears that the anhydrite and the related dolomite are younger than the carbonate. Nowhere is there evidence that the carbonate lithofacies grade laterally into the massive anhydrite.

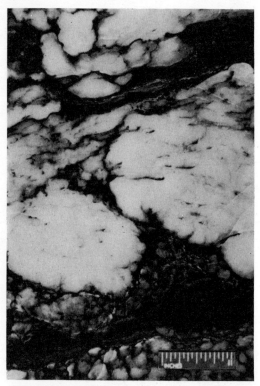

Fig. 27.—Massive nodular anhydrite. Nodules are less densely packed than in Fig. 26, but are separated by thin dolomite streaks and organic laminae. The compaction or "wrapping" effect of dolomite around anhydrite nodules is very noticeable. Texaco No. 3 G. D. Fehr Government, polished core, depth 5,404 ft.

Much has been written recently concerning the depositional relations of dolomite, gypsum, and anhydrite in Recent sediments in intertidal and supratidal environments along coastlines in arid or evaporative climates (Illing et al., 1965; Deffeyes et al., 1965; and Kinsman, 1964, 1965). It appears that the formation of these dolomite muds in the supratidal environment is "penecontemporaneous," by the reflux of hypersaline waters (Adams and Rhodes, 1960; Deffeyes et al., 1965); the high-magnesium water moves downward through the carbonate sediments of the supratidal flat to form the dolomite. In order to increase the Mg/Ca ratio of the hypersaline waters, gypsum and/or anhydrite must first be precipitated (Illing et al., 1965). Talmage and Wootton (1937) and others have pointed out that gypsum can also form by evaporation of groundwater in the vadose or capillary zone. The net result is that the gypsum and/or anhydrite will de-

velop at the expense of the soft sediments and may eventually replace them.

Deffeyes et al. (1965, p. 87) pointed out that the only environmental requirements are a seasonally or permanently dry climate so that evaporation may exceed precipitation, and a nearly flat sediment surface near sea level to provide a supratidal environment of sufficient areal extent. According to R. C. Murray (1966, personal commun.), ancient massive nodular anhydrite represents a supratidal environment, and the enclosing tidal-flat dolomite indicates an intertidal to supratidal environment. Therefore, the thick, massive anhydrite and related dolomite in the Ismay zone in the Cache field would have been deposited under similar conditions.

RELATION OF NORMAL-MARINE BUILDUP AND EVAPORITIC CYCLES

Much of the thickening of section in the Cache field is due primarily to the formation of the algal calcirudite breccias, indicating that the mounds grew at a faster rate than the surrounding areas. Stratigraphic relations show that the thinner deposits of the flanking wells are equivalent for the most part to the thicker deposits of the mound wells. As the algal buildups became "stacked" one on another, vertically, the areal extent of each succeeding mound was reduced.

The initiation of the evaporitic cycle may have occurred before the deposition of the final mound buildup of the normal-marine buildup cycle. As the Ismay seas began to recede, the mound buildup may have been exposed as a low hill on a nearly flat sediment surface. "Evaporitic" dolomite was deposited and gypsum and/or anhydrite crystals grew in the soft sediments on a supratidal flat. Dolomite containing algal-mat laminations formed in intertidal flats. The gypsum and anhydrite formed at or near the surface at the expense of the dolomitic mudstone and eventually increased in thickness as the dolomitic mudstone was displaced or squeezed out. Finally, minor transgressions of the sea crossed the region and caused deposition of the uppermost carbonate units at the end of Ismay time.

RELATION OF OIL TO POROSITY AND PERMEABILITY

The main productive reservoir rocks in the Cache field are in zones of favorable porosity and permeability within the algal mounds (algal calcirudite breccia) and the dolomitized shelly mudstone. Minor production comes from "earthy" and "evaporitic" dolomites and dolomitized zones in the foraminiferal-pelletal limestone.

Porosity in the algal calcirudite breccia is mostly secondary, although some interparticle voids may be primary. Secondary porosity is due primarily to the leaching of the phylloid (*Ivanovia*) algae and the formation of vugs and voids. Where secondary leaching of the algae has occurred, porosity is more than 14 percent, and permeability may exceed 450 md. Vugs and voids may be caused in part by collapse brecciation of the mud intraclasts when the algae was leached.

Porosity in the shelly mudstone is secondary and is best developed where the fossils have been completely leached and the mudstone has been dolomitized. Well-developed intercrystalline porosity is common where the crystals are widely spaced. Within the productive zones, the dolomitized shelly mudstone has a porosity of 10–15 percent and permeability of 15–20 md.

Diagenetic processes have caused most of the porosity and permeability in rocks in the study area; however, some of these processes, such as (1) recrystallization by calcite, (2) cementation by calcite, and (3) infilling of voids and replacement of bioclastic material by anhydrite, have greatly reduced the porosity and permeability. Most of these diagenetic processes appear to have taken place shortly after deposition of the sediments.

CONCLUSIONS

The Pennsylvanian Ismay zone of the Paradox Formation in the Cache field is represented by sediments deposited in two different but similar sedimentary environments. The earlier one, a normal-marine environment, involved intertidal and normal-marine processes similar to modern-day carbonate deposition of the Florida Bay-Bahama area. The later one, an evaporitic environment, involved supratidal and intertidal processes similar to modern-day sedimentation in the Trucial Coast, Persian Gulf area.

Lithofacies of the normal-marine cycle, although diverse in mineral composition, the type of skeletal and nonskeletal grains, texture, and structure, represent sedimentation in a somewhat restricted, widespread, shallow sea. The gross pattern of the facies suggests that the wave energy was minimal, that the waters ranged from slightly

stagnant to well oxygenated, and that the salinity was normal to somewhat higher than normal.

Cyclic changes in relative sea level probably produced the various changes in the lithofacies, but the rapid outward and upward growth of the algal mounds significantly influenced the sedimentational and biological pattern around them. Most facies could have been, and probably were, deposited simultaneously as a result of lateral variations in the factors which affect sedimentation and organic growth.

Lithofacies of the evaporitic cycle are characterized by extensive dolomitization, associated massive nodular anhydrite, lack of fauna, thin organic laminae, and fine-grained carbonate rock. These features indicate sedimentation of "penecontemporaneous" dolomite muds on a supratidal flat. The formation of the massive nodular anhydrite and replacement dolomite probably occurred on an almost flat surface near sea level along an arid coastline by downward-flowing hypersaline waters and/or by evaporation of groundwater in the vadose zone.

SELECTED BIBLIOGRAPHY

Adams, J. E., and M. L. Rhodes, 1960, Dolomitization by seepage refluxion: Am. Assoc. Petroleum Geologists Bull., v. 44, no. 12, p. 1912–1920.

Baars, D. L., J. W. Parker, and J. Chronic, 1967, Revised stratigraphic nomenclature of Pennsylvanian System, Paradox basin: Am. Assoc. Petroleum Geologists Bull., v. 51, no. 3, p. 393–403.

Choquette, P. W., and J. D. Traut, 1963, Pennsylvanian carbonate reservoirs, Ismay field, Utah and Colorado, *in* Shelf carbonates of the Paradox basin: Four Corners Geol. Soc., p. 157–184.

Deffeyes, K. S., F. J. Lucia, and P. K. Weyl, 1965, Dolomitization of Recent and Plio-Pleistocene sediments by marine evaporite waters on Bonaire, Netherland Antilles, *in* Dolomitization and limestone diagenesis—a symposium: Soc. Econ. Paleontologists and Mineralogists Spec. Publ. 13, p. 71–88.

Dunham, R. J., 1962, Classification of carbonate rocks according to depositional texture, *in* Classification of carbonate rocks—a symposium: Am. Assoc. Petroleum Geologists Mem. 1, p. 108–121.

Elias, G. K., 1963, Habitat of Pennsylvanian algal bioherms, Four Corners area, *in* Shelf carbonates of the Paradox basin: Four Corners Geol. Soc., p. 185–203.

Herman, G. and C. A. Barkell, 1957, Pennsylvanian stratigraphy and productive zones, Paradox Salt basin: Am. Assoc. Petroleum Geologists Bull., v. 41, no. 5, p. 861–881.

Illing, L. V., A. J. Wells, and J. C. M. Taylor, 1965, Penecontemporary dolomite in the Persian Gulf, *in* Dolomitization and limestone diagenesis—a symposium: Soc. Econ. Paleontologists and Mineralogists Spec. Publ. 13, p. 89–111.

Khvorova, I. V., 1946, On a new genus of algae from the middle Carboniferous deposits of the Moscow basin: Akad. Nauk U.S.S.R. Doklady, v. 53, no. 8, p. 737–739 (in Russian).

Kinsman, D. J. J., 1964, The Recent carbonate sediments near Halat el Bahrani, Trucial Coast, Persian Gulf, *in* Developments in sedimentology, v. 1: Amsterdam, Elsevier, p. 185–192.

—— 1965, Gypsum and anhydrite of Recent age, Trucial Coast, Persian Gulf, *in* Proceedings of second salt symposium: Northern Ohio Geol. Soc., p. 302–326.

Klement, K. W., 1966, Studies of the ecological distribution of lime-secreting and sediment-trapping algae in reefs and associated environments: Neues Jahrb. Geologie u. Paläontologische Abh., v. 125, p. 363–381.

—— 1967, Practical classification of reefs and banks, bioherms and biostromes (abs.): Am. Assoc. Petroleum Geologists Bull., v. 51, no. 1, p. 167 –168.

Malin, W. J., 1958, A preliminary informal system of nomenclature for a part of the Pennsylvanian of the Paradox basin, *in* Intermtn. Assoc. Petroleum Geologists Guidebook, 9th Ann. Field Conf., p. 135–137.

Merriam, D. F., 1962, Late Paleozoic limestone "build-ups" in Kansas, *in* Geoeconomics of the Pennsylvanian marine banks in southeast Kansas: Kansas Geol. Soc., 27th Field Conf., p. 73–81.

Murray, R. C., 1964, Origin and diagenesis of gypsum and anhydrite: Jour. Sed. Petrology, v. 34, p. 512–523.

Peterson, J. A., 1959, Petroleum geology of the Four Corners area: Proc. 5th World Petroleum Congr., sec. 1, paper 27, p. 499–523.

—— 1966, Stratigraphic *vs.* structural controls on carbonate-mound hydrocarbon accumulation, Aneth area, Paradox basin: Am. Assoc. Petroleum Geologists Bull., v. 50, no. 10, p. 2068–2081.

—— and H. R. Ohlen, 1963, Pennsylvanian shelf carbonates, Paradox basin, *in* Shelf carbonates of the Paradox basin: Four Corners Geol. Soc., p. 65–79.

Pray, L. C., and J. L. Wray, 1963, Porous algal facies (Pennsylvanian), Honaker Trail, San Juan Canyon, Utah, *in* Shelf carbonates of the Paradox basin: Four Corners Geol. Soc., p. 204–234.

Talmage, S. B., and T. P. Wootton, 1937, The nonmetallic mineral resources of New Mexico and their economic features: New Mexico Sch. Mines Bull. 12, 159 p.

Wengerd, S. A., and M. L. Matheny, 1958, Pennsylvanian System of Four Corners region: Am. Assoc. Petroleum Geologists Bull., v. 42, no. 9, p. 2048–2106.

Copyright © 1973 by The American
Association of Petroleum Geologists

The American Association of Petroleum Geologists Bulletin
V. 57, No. 6 (June 1973), P. 1053-1074, 17 Figs., 2 Tables

Rock and Biotic Facies Associated with Middle Pennsylvanian (Desmoinesian) Algal Buildup, Nena Lucia Field, Nolan County, Texas[1]

DONALD FRANCIS TOOMEY[2] and H. DALE WINLAND[3]
Tulsa, Oklahoma 74102

Abstract Six distinctive limestone facies, plus shale and sandstone, have been defined by a detailed petrologic and paleontologic study of five cores through a Middle Pennsylvanian (Desmoinesian) carbonate buildup in the Nena Lucia field area, Nolan County, Texas. The limestone facies, designated according to their most outstanding characteristic(s) are: (1) crinoidal, (2) pelletal-foraminiferal, (3) algal-plate, (4) algal-intraclast, (5) intraclastic, and (6) micritic. The first four facies display a restricted distribution pattern relative to the carbonate buildup, whereas the last two facies may be present in any position within the carbonate sequence. Quantitative limits have been delineated for the characteristic petrographic parameters of each facies present within the line of cross section through Nena Lucia field. Subtle petrographic and paleontologic differences in the sandstone and shale from the northwest (front) side and southeast (back) side of the bank are present in the studied cores. Distinct vertical paleontologic changes corroborate and strengthen the petrographic criteria. In addition, certain paleontologic variables appear to show subtle lateral changes that may be applicable in attempting to determine relative position on, or proximity to, the algal-plate mound buildups.

Introduction

Pennsylvanian rocks of late Desmoinesian, Missourian, and Virgilian ages in West Texas undergo complex facies changes in passing from platform limestones, sandstones, and shales on the Eastern shelf to basinal shales and sandstones westward into the Midland basin (Fig. 1). The Pennsylvanian section is transgressive, with progressively younger basinal sediments overlying shelf sediments in an eastward direction. Of special interest among the facies changes is a series of north-striking, elongate limestone buildups, mounds, or "island-type" banks near the basin-platform margin. These banks did not form a continuous barrier in time and space, because there is no abrupt change in the shales from the front to the back of the buildups. At most places, the Caddo Formation (lower Desmoinesian) served as the base on which the banks grew. The Caddo consists of oolitic, intraclastic, and skeletal calcarenites representing shelf facies.

Some of the buildups along the Eastern shelf are restricted to the late Desmoinesian interval; others culminate in rocks of Missourian or Virgilian age. Some buildups are relatively low, subtle features a few hundred feet thick; others may

swell to a thickness of more than 1,000 ft within 1 mi. The buildups are encased within dark-gray shale of Pennsylvanian age, and as such they afford excellent traps for petroleum. Many of these carbonate banks produce oil and gas in economic quantities at relatively shallow depths, ranging from 5,000 to 7,000 ft.

Nena Lucia field is close to the northern limits of a chain of related limestone buildups that extends from northern Nolan County southward to southern Schleicher County, a distance of approximately 125 mi. The field is irregularly shaped, elongate, and trends northeast-southwest. It is approximately 11 mi long and 1-3 mi wide (Fig. 2). The producing porous algal-plate zone is up to 100 ft thick. The field was discovered by C. L. Norsworthy, Jr.'s, No. 1 Helen Compton "A" in 1955, and has produced approximately 26.5 million bbl of petroleum. Estimates of ultimate field recovery range up to 40 million bbl.

Exploratory drilling for island-type limestone banks or buildups has met with limited success, inasmuch as there is no known method for predicting the nearness to a bank, or the direction to one from a previously drilled well. Strata penetrated in wells in a platform direction from a buildup have much the same gross appearance in both sample and mechanical log characteristics as those basinward from the buildup. Accord-

© 1973. The American Association of Petroleum Geologists. All rights reserved.

[1]Manuscript received, October 10, 1972; revised and accepted, November 28, 1972.

[2]Amoco Production Company, Research Center. Present address: Faculty of Earth Science, University of the Permian Basin, Odessa, Texas 79762.

[3]Amoco Production Company, Research Center.

The writers are indebted to the management of Amoco Production Company for the release of this material. The work for this report was done in 1964 as an integrated company project. We also appreciate the help given to us by the following division operating personnel: Ben Baldwin, Ray Stotler, and Frank Horton. We are especially grateful to Frank Horton for aiding in obtaining core material and acting in the capacity of liaison between the Research Center and the operating division. Harvey J. Meyer was a great help in performing the statistical analyses of our raw data, and George Sanderson furnished the necessary fusulinid age-dating framework. We also profited by discussions with Karl W. Klement on the role of phylloid algae in mound genesis.

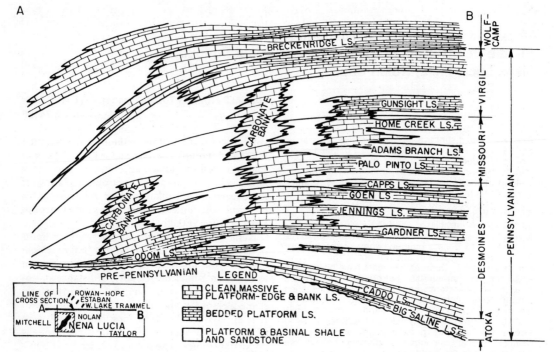

Fig. 1—Schematic east-west cross section of late Paleozoic sediments from Eastern shelf to Midland basin.

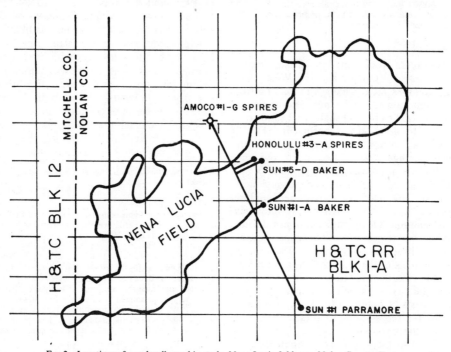

Fig. 2—Locations of cored wells used in study, Nena Lucia field area, Nolan County, Texas.

ingly, even if the flank side of a buildup is penetrated, no basic information has been available to determine in which direction to prospect for a second offset or step-out well.

The purpose of this study was to gain a detailed understanding of the petrographic and paleontologic parameters of the sediments in and associated with a buildup; of their interrelation and intrarelations which, it is thought, will reflect the depositional environments; and of changes in the distribution of environments with time. It was hoped this knowledge would facilitate prediction of nearness to and direction toward "island-type" carbonate buildups from present well control in this area, as well as in other areas with similar Pennsylvanian carbonate buildups.

Harrington and Hazlewood (1962), in a philosophical study comparing recent Bahamian landforms with the so-called Nena Lucia dune-reef-knolls, inferred an eolian depositional provenance for Nena Lucia, an inference not supported by the results of this study. Later, Kerr (1969) briefly noted that Nena Lucia produces from massive phylloid algal limestone of Desmoinesian (Stawn) age. He correctly stated that these algae have a significant bearing on the quality and, in places, the existence of the reservoir. In discussing and summarizing the attributes of phylloid algal reservoirs, he stated that they "commonly are surrounded by nonporous mudstone and wackestone and thus fall in the class of reservoirs wherein sediment genesis is an important factor in pore origin. An initial pore network controlled by plate morphology helps localize later diagenetic events, which ultimately produce a well-connected, predominantly large-pore network."

These features are well shown by the evolution of the Nena Lucia phylloid algal buildup.

PROCEDURE

A detailed petrographic and paleontologic study of the limestone facies present in representative cores from the Nena Lucia field was completed. Computer methods for analyzing the vast amount of multivariate data obtained were applied in each phase of the study. A supplementary study aimed at establishing fusulinid time correlations within the cored section also was completed.

After an initial examination of available geologic reports, sample logs, and electric logs in the Nena Lucia area, five wells representative of the front (northwest) side, the back (southeast) side, and the buildup (or mound) proper were chosen for detailed investigation (Fig. 2). It was believed that all facies present within the model area

would be represented in this group of well cores. Full cores were available through the major part of the limestone-buildup section in the selected wells.

The procedure was as follows: (1) a 1-in. slab was cut along the side of each core and preserved; (2) the remainder of the core was laid out to form a continuous section and examined under a binocular microscope; from this examination a generalized description of the cored interval was prepared and the section was subdivided into lithologic units; (3) petrographic and paleontologic samples were taken from the core simultaneously so that features to be studied by one method would not be destroyed in sampling for the other; (4) two horizontal, disc-shaped slices of core, each approximately 2 in. thick, were cut from the core about every 2 ft; (5) the remainder of the core was broken into thin plates with a hydraulic rock crusher in a search for megafossils; the megafossils were processed, identified at the generic level, and catalogued; (6) samples for fusulinid age dating were taken from the broken core, and later thin sectioned, identified, and catalogued; (7) the first of each pair of 2-in. slices was crushed and dissolved in formic acid to isolate agglutinated Foraminifera, conodonts, and silicified organic debris; the acid residue then was examined under a binocular microscope and all organisms were picked and mounted on paleo slides, identified at the generic level, counted, and tabulated; (8) the second of each pair of 2-in. core slices was cut into segments and processed for lithologic studies; a thin section and a polished surface were prepared from each segment; thin sections and polished slabs were used to study and delineate the major lithologic parameters; estimates were made on each sample to determine whether the rock had a matrix or grain-supported texture, the amounts and kinds of matrix (sparry or micritic), major grain types (intraclasts, pellets, skeletal grains, and oolites), skeletal grains (echinoderms, bryozoans, brachiopods, foraminifers, algae, corals, ostracods, mollusks), impurities (quartz sand, argillaceous material, pyrite, chert, glauconite), and secondary features (recrystallization, dolomitization, porosity, oil stain); a third segment from the 2-in. slice was powdered and used in X-ray mineralogic studies; the percentage of dolomite was determined from this sample. Porosity determinations from conventional core analyses had been made previously and were available from most of the cored interval.

As each well study was completed, all the data were tabulated and plotted in chart form for visual analysis. The core interval then was subdi-

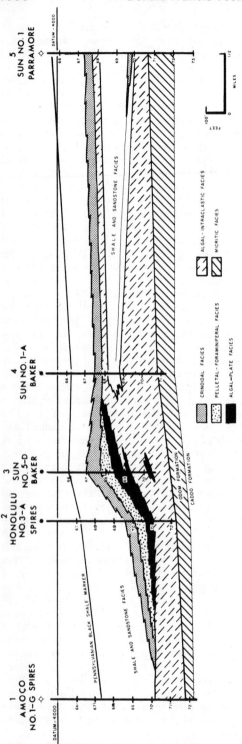

Fig. 3—Northwest-southeast cross section across Nena Lucia field showing delineated facies relations. Line of cross section is shown in Figure 2.

vided into lithologic units on the basis of total composition; these units proved to be facies of one another. Boundaries between the units (or facies) were arbitrarily picked by utilizing both petrologic and paleontologic data, inasmuch as most units are either gradational or interfingering. All data were then entered on punch cards and processed by computer methods with the use of factor or vector analysis and other programs. The average and range of composition of facies, the affinity of samples toward the various facies, the relation of one factor to another, the relation between samples, and changes in facies between wells were studied with these programs.

To determine whether the critical lithologic data could be obtained from drill cuttings, a synthetic set of "cuttings" was made from the Honolulu No. 3-A Spires core. This was done by taking a chip every foot and compositing the chips in 10-ft intervals. These composite samples were crushed in a jaw crusher and screened to remove any pieces larger than 1/4 in. in diameter. The size of samples obtained in this manner was judged comparable to the best samples obtained from wells drilled with good mud control. The "samples" were examined under a binocular microscope and estimates were made of the relative abundance of those factors thought essential for facies identification. The gross lithologic picture of the penetrated section obtained by this procedure closely resembled the analysis obtained by detailed core work. It was concluded that the information necessary for gross facies identification can be obtained from detailed examination of bit cuttings. The size of individual cuttings need not be as coarse as those used in our investigation. The minimum size that can be used effectively is about 1/8 in. in diameter.

PETROGRAPHIC AND PALEONTOLOGIC
DESCRIPTION OF FACIES

Six different limestone facies, plus shale and sandstone, have been defined by the described procedure in the upper Desmoinesian section of the Nena Lucia area. These six facies, named according to their distinctive component(s) are: (1) crinoidal, (2) pelletal-foraminiferal, (3) algal-plate, (4) algal-intraclast, (5) intraclastic, and (6) micritic. The first four facies apparently have restricted distribution in relation to the carbonate bank, and the latter two may be present anywhere within the limestone sequence. The writers' interpretation of the distribution and relations of facies between the wells studied is shown in Figure 3. The distribution of only the main bodies of the facies are shown; thin interbeds of shale and intraclastic and micritic limestone facies were

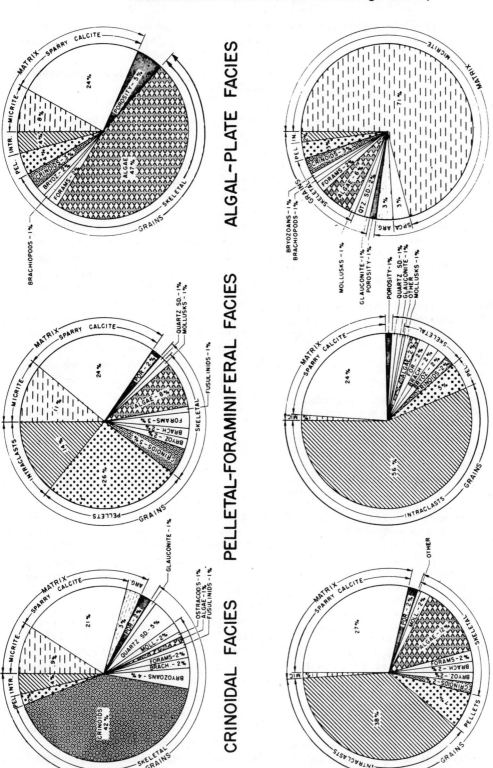

FIG. 4—Average composition of Nena Lucia facies, and range in composition of distinctive variables

omitted in this figure. The salient petrographic features of the facies are shown graphically on Figure 4. Photographs of polished core slabs and thin sections on Figures 5 and 6 show the typical lithologic character of each limestone facies. All fossils discussed in the following description of facies are illustrated in photographs appearing on Figures 7-12.

Distinct vertical petrologic and paleontologic changes were noted within the cored interval in the Nena Lucia field. The vertical changes are pronounced and allow recognition of a vertical sequence of rocks and biota. These rock and biotic subdivisions can be recognized laterally along the line of cross section (see Fig. 3). Nevertheless, the individual lithologies appear to be of uniform gross aspect laterally, and lithologic patterns do not show any pronounced variation which would allow accurate delineation of relative position on, or in proximity to, a mound buildup. In contrast, the paleontologic data appear to show subtle lateral variations which may be applicable in determining position relative to our drilling target, the algal-plate mound buildup. Accordingly, a detailed discussion of each of the recognizable facies is given.

Crinoidal Facies

The crinoidal facies in all the wells studied is characterized by an abundance of skeletal grains, of which crinoidal debris is the dominant element. Typical examples of crinoidal limestones are shown on Figures 5A and 6B. The facies is unique in the relatively high content of very fine quartz-sand grains and glauconite pellets. Consistently associated with the crinoidal debris are fragments of bryozoans and brachiopods. The rock is grain supported and typically cemented with sparry calcite. This rock may be called a crinoidal calcarenite. Grain size of the particulate carbonate matter ranges from silt-size fragments to crinoidal columnals up to 1 in. in diameter. Fine-grained crinoidal limestones generally are darker and more micritic than the coarse-grained varieties. In the five cores studied no oil staining was noticed in rocks of this facies, and most porosities through crinoidal intervals are low.

The crinoidal facies is most commonly distributed along the top and in front of the limestone banks. In addition, thin stringers of crinoidal debris are interbedded with the overlying shale and sandstone and, in places, may be interbedded with the upper part of the underlying pelletal facies.

The relative abundance of paleontologic components for this facies is shown graphically on Figure 13. A varied biota is present; in fact, this is the only facies in which such elements as brachiopods and conodonts are well represented. It also is the only facies in which relatively diverse calcareous and agglutinated foraminiferal assemblages occur together in moderate abundance.

Of the megafauna (Figs. 11, 12) the most abundant and common brachiopod in this facies is the productid *Kozlowskia;* next in abundance is the small spiriferacid *Phricodothyris.* Both of these forms show some decrease in abundance toward the shelf area. A few of the brachiopods appear to show some lateral restriction; *i.e., Composita* and the spiriferid *Neospirifer* commonly are found in the wells on the mound buildup, with *Composita* more common along the outer edge of the buildup. The chonetid brachiopod *Chonetinella* and the inarticulate brachiopods have a somewhat wider distribution.

Rugose corals appear to be more common in the two wells that penetrated the mound buildup, although they do extend onto the shelf area.

The foraminifers are well represented in this facies, with encrusting types as the most abundant forms. Encrusting genera such as *Tuberitina* (Fig. 9C-E), *Tetrataxis* (Fig. 8A-B), and *Ammovertella* (Fig. 10D) appear to be distributed through this facies with little change in relative abundance. The encrusting forms *Apterrinella* and *Hedraites* (Fig. 9J-L) are most abundant in the backmound area and decrease sharply as the shelf area is approached.

The mobile foraminifers appear to have distinct lateral restriction and significance although, when the mobile forms are plotted together, the overall pattern is one of increasing abundance toward the shelf area. Specimens of *Endothyra* (Fig. 9H) are most abundant in the mound buildup, whereas fusulinids are relatively common along the entire trend, together with *Climacammina* (Fig. 9B). *Bradyina* (Fig. 9I) and *Globivalvulina* (Fig. 8C) are most abundant in the shelf area and are sparsely distributed throughout the rest of the facies.·

Siliceous sponge spicules of various shapes and sizes (Fig. 10M-O) are common in the acid residues of crinoidal facies samples from the backmound area and on the shelf.

In general, the total aspect of the biota suggests deposition in a relatively shallow-water marine

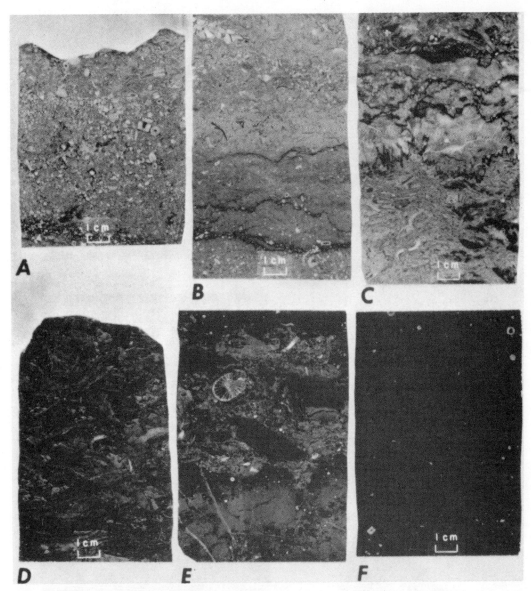

Fig. 5—Photographs of polished core slabs showing major Nena Lucia rock facies. **A.** Moderately sorted crinoidal packstone with grains ranging in size from very fine to medium sand. Particulate matter is predominantly disarticulated crinoid columnals. Some gray micritic intraclasts are present, especially near top of slab. Dark, intergranular areas are clear sparry calcite; stylolite near bottom of photograph. Sun No. 5D Baker, crinoidal facies (6,787 ft.). **B.** Pelletal limestone with thin intraclastic interbeds. Pelletal parts are tan and gray vaguely granular streaks; small grains associated with streaks are crinoid columnals. Intraclastic zones across central part of slab contain concentrations of discrete particles up to 1 cm in diameter separated by sparry cement or pelletal limestone. Dark streaks in lower half of slab are shale concentrations along stylolitic partings. Sun No. 5D Baker, pelletal-foraminiferal facies (6,798 ft). **C.** Algal-plate packstone with sparry and micritic matrix. Dark micritic matrix appears to be trapped in pockets bound by areas of algal plate. Most algal plates are fragmented. Note stylolites and small, very scattered, crinoid columnals. Sun No. 5D Baker, algal plate facies (6,890 ft). **D.** Intraclastic algal-plate packstone with intraclast and phylloid algal debris set in dark matrix. Particulate matter poorly sorted with intraclasts ranging up to 1 in. in maximum dimension. Sun No. 1A Baker, algal-intraclast facies (6,985 ft). **E.** Intraclastic packstone with prominent large intraclasts set in micritic matrix. Note conspicuous rugose coral. Large intraclast at bottom appears to show internal disruption. Intraclastic facies, Amoco 1G Spires (7,018) ft). **F.** Matrix-supported silty mudstone with less than 10 percent skeletal grains. Micritic facies, Amoco 1G Spires (7,181 ft).

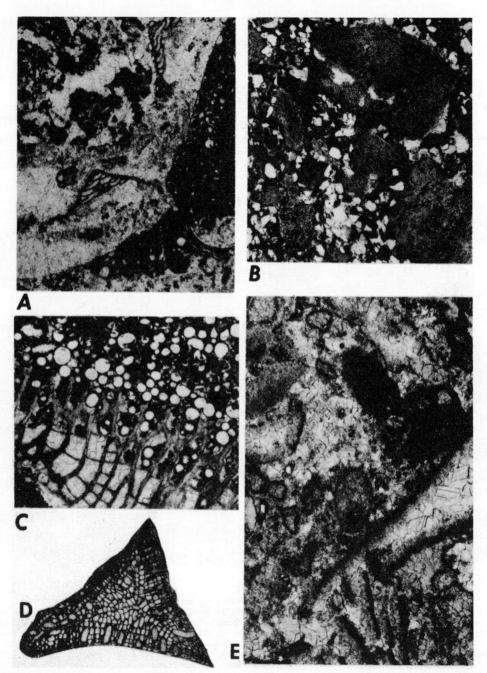

Fig. 6—Thin section photomicrographs showing characteristic Nena Lucia microfacies, all X20. **A.** Recrystallized algal plates encrusted by foraminifera *Tetrataxis* and other encrusting foraminifers, Sun No. 1 Parramore, algal-intraclast facies (6,780 ft). **B.** Silty/pelletoidal crinoidal packstone, Honolulu No. 3A Spires, crinoidal facies (6,970 ft). **C.** Tabulate coral *Chaetetes*. Circular "calcispheres?" abundant in upper half, Amoco 1G Spires, algal-intraclast facies (7,074 ft). **D.** Triangular-shaped prismoporid bryozoan, Honolulu No. 2 Spires, pelletal-foraminiferal facies (6,869 ft). **E.** Recrystallized and broken algal plates (dark edge outlines algal plate and pelletoids and/or intraclasts), Sun 5D Baker, algal-plate facies (6,864 ft).

environment of normal salinity. That appreciable hydraulic energy was expended against the bottom sediments and associated faunal elements is shown by the abrasion and fragmentation of the megafauna. The clastic quartz silt content decreases from top to bottom of the facies. The crinoidal facies appears to have capped the bank, inasmuch as this facies occupies the most seaward position within the facies assemblage and also forms the lag deposits over all older sediments as the facies assemblage migrated eastward.

Pelletal-Foraminiferal Facies

Petrographically, this facies is characterized by the abundance of silt-size pelletoids composed of structureless cryptocrystalline calcite (micrite), and micritic intraclasts. Inasmuch as both allochems are formed of the same material, the pelletoids probably were derived through fragmentation and abrasion of original larger intraclasts, and are not fecal in nature. The intraclasts probably were formed by wave current disruption of the bottom sediment during periods of intermittent storm action.

A typical example of pelletal-foraminiferal limestone is shown on Figure 5B. At magnification normal for sample analysis (9x), the rock may look micritic and structureless, but abundant minute dark and light specks give a clue to the true pelletal nature. Recrystallization and porosity development within the facies are scattered. At places in Nena Lucia field porous streaks in the pelletal-foraminiferal facies are productive.

In the five cores studied, rocks of this facies lie below the crinoidal facies in the area of carbonate buildup and are associated closely with the underlying algal-plate facies.

The paleontologic components of this facies have been tabulated, and some of the more meaningful elements plotted on Figure 14. Faunally, this facies contains an abundant and diverse suite of calcareous foraminifers, both mobile and encrusting types. Fusulinids are most numerous, and their distribution pattern shows greatest abundance on the front of the mound buildup, with an apparent decrease in total number toward the back of the buildup. The relatively large mobile foraminifer *Bradyina* also has a similar distribution pattern, although on a much reduced scale. The foraminifers *Hyperammina* (Fig. 10I-J), *Minammodytes* (Fig. 10A), and *Endothyra* show an interesting reciprocal relation, in that they are distributed more abundantly on the back of the buildup. Significantly, the agglutinated foraminifer *Thurammina* (Fig. 10H) is present only on the back of the buildup.

Taxonomically, this facies contains the most diverse calcareous foraminiferal microfauna of all the facies present. It also contains a rather consistent component of algal debris other than sporadic phylloid algal plates. Small hollow spheres, herein termed "calcispheres" (Fig. 6C), are most common on the front of the buildup and are very scarce only on the back of the mound. Red algae, e.g., *Cuneiphycus* (Fig. 7C-D), are restricted to the front of the buildup, whereas the blue-green algae *Girvanella* (Fig. 7B) and *Tubiphytes* (Fig. 7F) are present on the back of the mound.

Of the megafauna, the rugose corals are not common on the front of the mound, but become more abundant on the back. With respect to the brachiopods, only three genera are represented, and these are somewhat limited in number. The productid *Kozlowskia* is present on the front of the mound buildup, whereas the relatively large productid *Linoproductus* (Fig. 11I) is restricted to the back. The small spiriferacid *Phricodothyris* (Fig. 11C-D) was found in both wells that penetrated this facies, but it is numerically more abundant in the well at the front of the mound.

Environmentally this facies seems to have been deposited in normal marine waters of shallow depth. Hydraulic energy gradients across the facies appear to be somewhat lower and less consistent than in the overlying crinoidal depositional environment. The restriction of the facies to the seaward side of the mound buildup suggests at least some controlling effect of the environment being exerted by the underlying mound itself.

Algal-Plate Facies

As the name indicates, rocks of this facies consist dominantly of phylloid algal material (Fig. 5C). This facies was present in two of the studied wells. The algal debris has been thoroughly replaced and recrystallized, and now appears as blades, "leaves," plates, or sinuous bands of sparry calcite outlined by micritic envelopes (Fig. 6E). The algal fronds are not in growth position and are packed together horizontally much like a pile of leaves. Although no whole specimens were found, neither are the fronds excessively fragmented. However, *in situ* brecciation of the algal plates is common.

The matrix is mostly sparry calcite, but numerous pockets and pods of micritic matrix are present among the plates. The sparry matrix is thought to represent both secondary infillings of original porosity and recrystallization of carbonate mud matrix. This is the only facies that is conspicuously recrystallized and porous throughout. Most of the production at Nena Lucia field

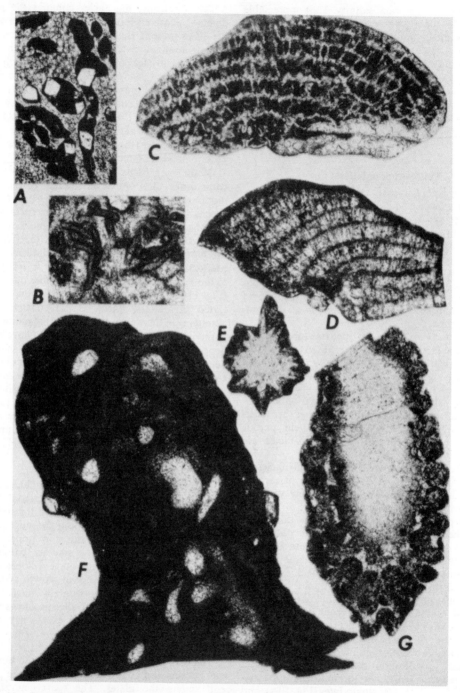

FIG. 7—Nena Lucia algal structures as seen in thin section; all X85; **A, C** from crinoidal facies, **B, E, G** from algal-intraclast facies, **D, F** from pelletal-foraminiferal facies. **A.** Algal? borings into crinoid columnal, Honolulu No. 3A Spires (6,998 ft). **B.** Tubules of blue-green alga *Girvanella,* Sun No. 5D Baker (6,993 ft). **C** and **D.** Thin-section cuts of red alga *Cuneiphycus.* **C,** Sun No. 1 Parramore (6,791 ft); **D,** Honolulu No. 3A Spires (7,038 ft). **E.** Polyactinal sponge spicule circumscribed by micrite envelope, Sun No. 1 Parramore (6,788 ft). **F.** Blue-green alga? *Tubiphytes,* Sun No. 5D Baker (6,775 ft). **G.** Indeterminate dasyclad fragment, Sun No. 1 Parramore (7,034 ft).

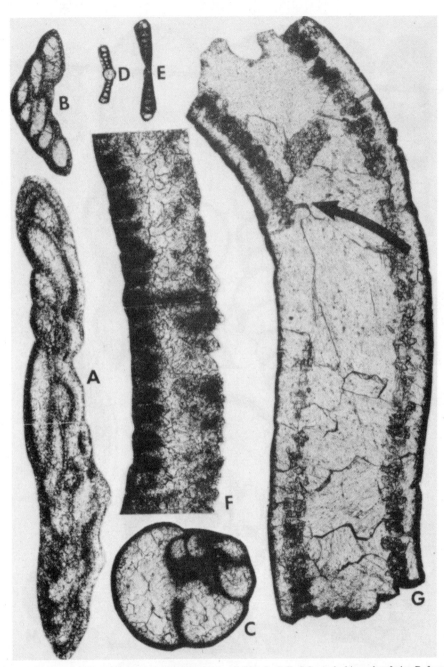

FIG. 8—Nena Lucia algae and foraminifers as seen in thin section, all X80; **A, C, E, G** from algal-intraclast facies, **B,** from pelletal-foraminiferal facies, **D,** from crinoidal facies, **F,** from algal-plate facies. **A** and **B.** Encrusting foraminifer *Tetrataxis*. Note large size of specimen **A** and morphologic adaptation for encrusting algal plates. Honolulu No. 3A Spires (**A,** 7,153 ft; **B,** 7,040 ft). **C.** Mobile foraminifer *Globivalvulina*, Sun No. 1 Parramore (6,775 ft). **D.** Mobile foraminifer *Eolasiodiscus*, Sun No. 1A Baker (6,776 ft). **E.** Mobile foraminifer *Eolasiodiscus*. Sun No. 1 Parramore (6,778 ft). **F.** Green (codiacean) algal plate *Ivanovia*. Note position of utricles along plate edge, Sun No. 1 Parramore (7,239 ft). **G.** Green (codiacean) algal plate *Eugonophyllum* (arrow points toward utricles), Sun No. 5D Baker (7,005 ft).

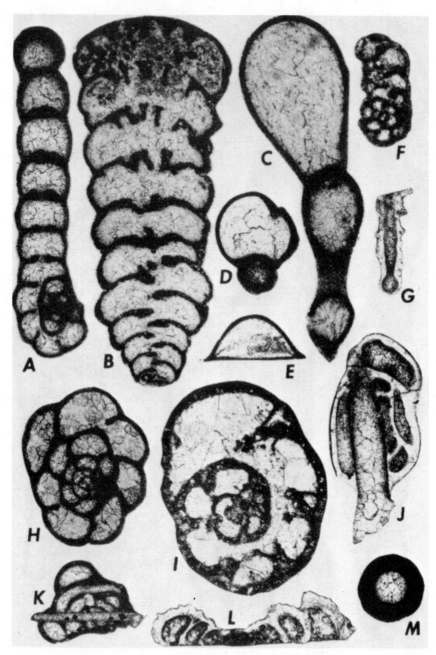

Fig. 9—Representative Nena Lucia calcareous foraminifers as seen in thin section, all X80 unless otherwise indicated. **A, H, J** from crinoidal facies; **G, I, M** from pelletal-foraminiferal facies; and **B-F, K-L** from algal-intraclast facies. **A.** Mobile foraminifer *Endothyranella,* Sun No. 5D Baker (6,735 ft). **B.** Mobile foraminifer *Climacammina,* Sun No. 1 Parramore (6,778 ft). **C-E.** Thin section cuts of encrusting foraminifer *Tuberitina.* **C,** Sun No. 1A Baker (6,961 ft); **D,** Sun No. 5D Baker (6,998 ft); **E,** Sun No. 5D Baker (6,952 ft). **F.** Mobile foraminifer *"Palaeospiroplectammina,"* Sun No. 1 Parramore (6,788 ft). **G.** Mobile foraminifer *Syzrania,* Sun No. 5D Baker (6,795 ft). **H.** Mobile foraminifer *Endothyra,* Honolulu No. 3A Spires (6,974 ft). **I.** Mobile foraminifer *Bradyina,* Honolulu No. 3A Spires (7,053 ft); X32. **J-K.** Encrusting foraminifer *Apterrinella* (note encrusted object); Sun No. 5D Baker (**J,** 6,715 ft; **K,** 6,977 ft). **L.** Encrusting foraminifer *Hedraites,* Sun No. 1A Baker (7,038 ft). **M.** Algal? calcisphere, Sun No. 5D Baker (6,795 ft).

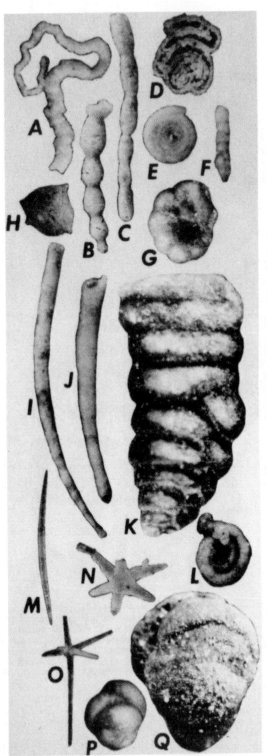

FIG. 10—Photomicrographs (X30) of representative Nena Lucia microfossils from acid residues and shale breakdown: A-C, H, K-L, P-Q are from crinoidal facies, all others from algal-intraclast facies. A. Encrusting agglutinated foraminifer *Minammodytes*, Sun No. 1A Baker (6,842 ft). B, C. Mobile agglutinated foraminifer *Reophax*, Sun No. 1A Baker (6,967 ft). D. Encrusting agglutinated foraminifer *Ammovertella*, Sun No. 1A Baker (6,809 ft). E. Mobile agglutinated foraminifer *Ammodiscus*, Sun No. 1A Baker (6,891 ft). F. Mobile agglutinated foraminifer *Bigenerina*, Sun No. 1A Baker (6,891 ft). G. Mobile calcareous foraminifer *Endothyra*, Sun No. 1 Parramore (6,778 ft). H. Mobile agglutinated foraminifer *Thurammina*, Sun No. 1A Baker. I, J. Agglutinated mobile foraminifer *Hyperammina*, Sun No. 1A Baker (6,889 ft). K. Calcareous mobile foraminifer *Climacammina*, Sun No. 1 Parramore (6,775 ft). L. Calcareous encrusting foraminifer *Apterrinella*, Sun No. 1A Baker (6,835 ft). M-O. Siliceous sponge spicules, Sun No. 1A Baker (6,888 ft). P. Calcareous mobile foraminifer *Globivalvulina*, Sun No. 1 Parramore (6,778 ft). Q. Calcareous mobile foraminifer *Bradyina*, Sun No. 1 Parramore (6,778 ft).

FIG. 11—Representative Nena Lucia megafauna; all X1 unless otherwise indicated; **A, F, K** from pelecypod-bearing shale interbedded with crinoidal facies; **B-E, J** from crinoidal facies; **G-I** from algal-plate facies. **A.** Pectinacid pelecypod *Streblochondria,* Sun No. 1A Baker (6,884 ft). **B.** Platycerid gastropod *Platyceras,* Honolulu No. 3A Spires (6,960 ft); X2. **C, D.** Spiriferacid brachiopod *Phricodothyris,* Honolulu No. 3A Spires (6,975 ft). **E.** Spiriferacid brachiopod *Amobocoelia,* Honolulu No. 3A Spires (6,975 ft); X3. **F.** Inarticulate brachioped *Orbiculoidea,* Sun No. 1A Baker (6,835 ft). **G, H.** Productid brachiopod *Kozlowskia,* Honolulu 3A Spires (7,081 ft); X2. **I.** Productid brachiopod *Linoproductus,* Honolulu 3A Spires (7,080 ft). **J.** Terebratulid brachiopod *Rhynchopora,* Honolulu No. 3A Spires (6,968 ft). **K.** Indeterminate productid brachiopod, Sun No. 1A Baker (6,877 ft).

Fig. 12—Representative Nena Lucia megafauna; X2 unless otherwise indicated; **A, B** from pelecypod bearing shale interbedded with crinoidal facies; **D** from algal-intraclast facies; **E-F** from algal-plate facies. **A, B.** Spiriferacid brachiopods, Sun No. 1A Baker (**A,** 6,818 ft; **B,** 6,875 ft). **C.** Chonetid brachiopod *Chonetinella*, Honolulu No. 3A Spires (6,987 ft). **D.** Trilobite pygidium *Ditomopyge*, Sun No. 1A Baker (6,925 ft); X3. **E, F.** Rostrospiracid brachiopod *Composita*, Honolulu No. 3A Spires (7,092 ft).

evidently comes from rocks of this facies. Porosity development in the rock results in part from solution of the organic components, but much is due to the development of extremely fine intercrystalline pores through recrystallization of the matrix material. Oil stain is common in rocks of this facies and thin seams of red iron stain have been observed.

This facies was found only within the carbonate bank proper in the wells studied. Evidently, prolific growth of the phylloid algae was responsible for the buildups.

The biota present in the algal-plate facies is shown graphically on Figure 15. First, it may be

noted that this facies has the lowest taxonomic diversity of all the carbonate facies in the Nena Lucia field. In total number the microfauna is dominated by the encrusting foraminifers. The calcareous encrusting form *Hedraites* is most abundant. The agglutinated encrusting foraminifers *Minammodytes* and *Ammovertella* appear to be restricted to the back of the buildup. The mobile agglutinated foraminifer *Hyperammina* also appears to be restricted to the back of the buildup. Calcareous mobile foraminifers such as *Climacammina* and *Bradyina* are uniform on both sides, whereas the Fusulinidae are most abundant along the front edge of the mound

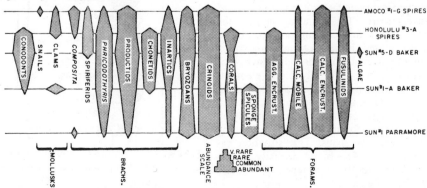

Fig. 13—Fossil distribution: crinoidal facies.

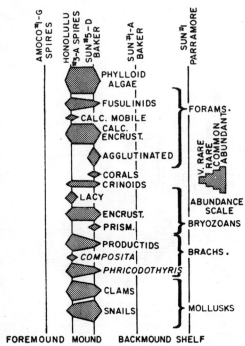

Fig. 14—Fossil distribution: pelletal-foraminiferal facies.

The growth habit of *Ivanovia* and *Eugonophyllum* cannot be established, as only broken and fragmented remains of leaflike bodies of the plant have been identified; neither attachment organs or complete "leaves" have been found. Nevertheless, there is some evidence that *Ivanovia* grew as an upright plant. This is based primarily on the bilateral symmetry of the internal structure of the plant—particularly the identical character of the pillar-like utricles of the outer layers of opposing walls—and the fact that the outer walls of the algal plate were carbonate encrusted. The explanation for the pronounced and prevalent diagenetic alteration (recrystallization) of the wall structure of the algal plates is obscure, but it may be related to its unstable mineralogy. In growth position the alga most probably was a few inches high and consisted of a rather limited number of relatively broad, somewhat rigid "leaves" or "fronds." These relatively broad leaves probably could not have withstood a high degree of current and wave turbulence; they probably thrived most prolifically in a zone somewhat below or beyond that of appreciable wave action. This alga is not the type that could have provided the successful framework or encrusting function on a growing "reef," though it is thought that this type was a very efficient sediment producer and sediment

buildup and are appreciably reduced in abundance at the back of the mound.

The bryozoans show an interesting pattern of distribution within this facies; fenestrate types are very common along the front of the mound buildup, whereas prismoporoid types (Fig. 6D) commonly are restricted to the back; the encrusting fistuliporoid types are relatively common throughout the facies.

Of the brachiopods, only two forms are present in any meaningful abundance. These are the productid *Kozlowskia* and the spiriferacid *Phricodothyris,* which are most abundant along the back of the buildup.

Within the algal-plate facies fossil invertebrates are relatively scarce. On the algal mounds the phylloid algae grew in such profusion, and occupied and dominated the sea-bottom living space, so that for all practical purposes most other organisms were excluded. The only biotic elements able to offer what might be considered faunal competition were epiphytic organisms. In this instance the ecologic niche was filled primarily by encrusting types of foraminifers and encrusting fistuliporoid-type byrozoans. This situation explains the relatively low biotic diversity in this facies compared with the other facies represented within the Nena Lucia field.

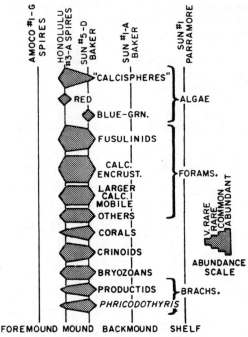

Fig. 15—Fossil distribution: algal-plate facies.

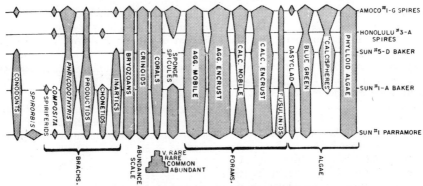

Fig. 16—Fossil distribution: algal-intraclast facies.

baffle and, in this sense, is probably analogous to the modern *Halimeda.*

As the mound built up to and above wave base and was intermittently subaerially exposed (as suggested by the red staining, recrystallization, and porosity development), it must have greatly affected depositional patterns around it. Probably the mound grew near or at the seaward margin of a shallow Pennsylvanian marine platform. Its specific location could be due to subtle features over which the late Desmoinesian seas originally transgressed. If these features became suitable for habitation by a phylloid alga during inundation, the plant, once established, would attempt to mold and maintain its living space in that particular environmental niche suitable to propagation of its kind during later relative sea-level changes. The vertical and horizontal extent of the organic deposit is the measure of the success of the plant complex.

Significantly, this is the uppermost facies in which pelecypods and gastropods are numerous. The pelecypods appear to be most abundant on the front of the mound buildup, whereas the gastropods are more common on the back.

Algal-Intraclast Facies

This facies, found in all the studied wells, is characterized by abundance of both intraclasts and skeletal debris (Fig. 5D). On polished surfaces the intraclasts appear as tan to gray, subangular to angular micritic inclusions in a fossiliferous sparry matrix. In thin section, the intraclasts appear as opaque to semiopaque masses of cryptocrystalline calcite. They are sparingly fossiliferous, although Foraminifera are a noticeable component. The principal component among the skeletal grains is algal debris. Most conspicuous of the minor skeletal constituents are encrusting foraminifers and bryozoans. Particulate matter making up the rock is very poorly sorted, and grain size within any specimen may range from silt-size skeletal debris to intraclasts several

inches in maximum diameter. Typically, the rock is "tight" and presents less evidence of recrystallization than does the algal plate facies.

In this study the algal-intraclast facies generally is found behind (southeast of) and beneath the algal-plate mounds, and has the widest areal distribution and the thickest development of all defined facies. Figure 4 shows the marked similarity of this facies to the intraclastic facies; the main difference is the greater amount of skeletal grains, especially algal debris, in the former. The algal-intraclast facies is regarded as a subfacies of the intraclastic facies but has been differentiated because of its unique distribution.

The paleontologic components of this facies, tabulated and graphically shown on Figure 16, show a marked increase in taxonomic diversity compared with the overlying algal-plate facies. The foraminiferal microfauna is dominated by encrusting types, both calcareous and agglutinated. Nevertheless, this is the only facies of those present in the Nena Lucia field that contains such an abundant and consistent component of agglutinated foraminifers, *i.e., Minammodytes, Ammovertella, Hyperammina,* and *Reophax.* In sheer numbers of specimens these foraminifers dominate the foraminiferal assemblage, and their distribution is persistent throughout the facies. The calcareous encrusting foraminifers *Hedraites* and *Tuberitina* also show a similar distribution pattern. The encrusting calcareous foraminifer *Apterrinella* is found only in the backmound and shelf areas. The mobile calcareous foraminifer *Globivalvulina* and the Fusulinidae occur only sparingly through most of the facies extent, but show marked increase in numbers as the shelf area is approached. Conversely, specimens of the genus *Endothyra* are most abundant in the foremound area.

Relatively well-preserved phylloid-algal remains, *Ivanovia* (Fig. 8F) and *Eugonophyllum* (Fig. 8G), are fairly common throughout this facies, as are tubules of the blue-green alga *Girvanella*. Algal calcispheres, thought to be algal spores, are rare to common in four cores, but absent within the shelf area.

Siliceous sponge spicules of various shapes are common in the backmound and shelf areas, very rare or absent in the mound buildup, but common in the foremound area.

The brachiopod megafauna is relatively diverse in number of taxa present, but specimens are not numerous. The only abundant forms are *Kozlowskia* and *Phricodothyris*, but fragments of a relatively large indeterminate productid (dictyoclostid-type) and fragments of inarticulate brachiopods *Composita*, *Chonetinella*, *Linoproductus*, and *Neospirifer* also are present. Most of these brachiopods are most abundant in either the foremound or backmound wells. Conodonts are more common in the backmound and shelf areas, and the marine worm *Spirorbis* is common only on the shelf.

Although the taxonomic diversity of this facies is relatively high, the number of individuals, with the marked exception of the agglutinated Foraminifera, tends to be low. The fauna could not be considered as having lived in other than a normal marine depositional environment, which suggests acceptable conditions of salinity and temperature. However, recent foraminiferal assemblages that are dominated by agglutinated genera are thought to represent lagoon-type or marginal environments of deposition. Also, modern occurrences of agglutinated foraminifers commonly are related to environments with low salinities, but in these cases the faunal diversity also is low although the total number of specimens may be high. Possibly, an area of periodic current activity was the milieu of this Pennsylvanian agglutinated foraminiferal assemblage. Some wave and current restriction may have been due in part to the presence of algal mounds in a seaward direction.

The intraclast formation process probably was accomplished within the depositional environment by storm and wave action tearing up a relatively consolidated sea bottom. However, the skeletal debris within this facies generally shows little wear or signs of abrasion, *i.e.*, the crinoidal debris is commonly in the form of relatively long articulated stems rather than individual ossicles.

Intraclastic Facies

The intraclastic facies consists almost entirely of intraclasts with little else (Fig. 5E). Whereas the algal-intraclast facies is relatively rich in skeletal grains (average 23 percent), this facies is marked by a paucity of organic debris (average 11 percent). The fossil elements usually reflect the fauna of the enclosing beds. The primary reason for differentiating both is to emphasize their respective distribution patterns. The algal-intraclast facies was found beneath and, for the most part, behind the algal mound, whereas rocks classified with the intraclastic facies may be found interbedded with all other facies. The fauna commonly reflects that of the enclosing beds but is restricted in number of individuals. The intraclastic facies probably represents a depositional environment quite similar to that of the algal-intraclastic facies.

Micritic Facies

The micritic facies, in marked contrast to all others, has a high content of carbonate (Fig. 5F). Rocks of this facies usually are matrix supported, whereas rocks belonging to all other facies are principally grain supported. Micritic limestones can be found interbedded with any other facies and characterize the Odom Formation, which underlies the algal-intraclast facies. Commonly, the Odom Formation is laminated. There is an increase downward (landward) in the amount of micrite interbedded with the algal-intraclast facies, and micrite is the dominant lithology in the Odom Formation underlying the algal-intraclast facies. The interfingering relation suggests concurrent deposition of the two lithic types.

Typically, rocks of this facies are dark, very dense, and little affected by recrystallization. In many places dark shale in thin stringers is associated with the micrite. These shales contain a relatively abundant pelecypod fauna and flood occurrences (extreme abundance) of calcareous mobile Foraminifera.

The faunal content of micritic-facies rocks interbedded with other facies mainly reflects the fauna of the enclosing beds. In those areas where micrite dominates the section, the fauna is severely restricted and only encrusting calcareous Foraminifera are common.

The micritic facies, especially the true micrites or mudstones, probably represents a depositional environment characterized by very quiet water. The carbonate mud constituting the micrite could have been deposited in deep water below wave base or in a highly restricted shallow-water environment. The absence of a marine fauna indicates an environment inhospitable for growth of marine organisms, and the interfingering relation with the algal-intraclast facies suggests that the two were deposited in juxtaposition. Thus, the micritic facies, especially the Odom, probably represents a depositional environment in extremely restricted, shallow water. If so, it would represent an end member in the pattern of deposition on a shallow marine shelf.

Shale and Sandstone Facies

Dark shales with interbedded sandstone stringers surround the limestone bank, and thin streaks of shale, 0.25-1.0 in. thick, are interbedded with the limestone. Subtle petrographic and paleontologic differences in the shale from the front and back sides of the bank are present in the cores. The shale in front of and over the bank is fissile and contains slightly calcareous and silty streaks. Limestone stringers interbedded with this shale generally are textureless micrite. The fauna in the shale is pyritized and is characterized by nektonic cephalopods. The position and relatively greater thickness of this shale and the type and sparsity of the fauna suggest a poorly oxygenated (basinal) type of depositional environment. In contrast, shale on the back of the bank is more calcareous, silty, and blocky. It contains a relatively greater proportion of scattered coarser skeletal grains, including crinoidal columnals, fenestrate bryozoans, pectinoid pelecypods (Fig. 11A), robust productid brachiopods, and some phosphatic inarticulate brachiopods. These faunal components suggest a lagoonal depositional environment for this type of shale.

Limestone stringers interbedded with the shale are commonly crinoidal calcarenite, but some beds of intraclastic limestone are present. In contrast to those crinoidal limestones deposited in front and above the bank, the stringers of crinoidal limestone in the back-bank shale contain no glauconite. The thin shale streaks interbedded in the limestone sequence are black and fissile, and contain an abundant microfauna consisting of flood occurrences of the calcareous Foraminifera *Globivalvulina*, *Bradyina*, *Climacammina* (mobile-types), and *Tetrataxis* (encrusting-type), with fewer *Endothyra* and fusulinids (mobile-types).

One thin shale stringer in the Sun No. 1 Parramore contained flood occurrences of the marine polychaete worm, *Spirorbis*. Because of the limited vertical extent of the stringers, little can be said concerning their depositional environment; they do suggest brief spans of geologic time when optimum conditions existed for the development of a "normal" robust microfauna.

Sandstones in the section generally are very fine grained and fossiliferous. Sandstones interbedded in the section directly overlying the carbonate bank are glauconitic, whereas those from back-bank shale are not. In the section just above the carbonate bank, interbedding of sandstone and crinoidal limestone can be observed. Commonly, evidence indicates that burrowing organisms have reworked the sediment.

FACIES SYNOPSIS

The limestones described in this report were deposited on a shallow marine shelf that covered broad areas of central Texas. This shelf extended northward through Oklahoma into the Mid-Continent, and was contemporaneous with a similar shelf that covered large sections of the western United States. The main elements influencing limestone deposition on this widespread shallow marine shelf were its extreme width, its relatively low regional dip, and the shallowness of the water over the shelf. All these factors acted in unison to (1) restrict water circulation, which in itself would cause variations in water salinity, temperature, clarity, and abundance of readily available nutrients; and (2) attenuate current and wave action, thus causing variations in the hydraulic energy acting upon the sediment surface. Bottom prominences may have influenced sedimentation locally by further restricting circulation.

The gross pattern of facies distribution in the Nena Lucia field suggests a decrease in hydraulic energy levels from northwest to southeast across the area studied; that is, decreasing from the relatively deeper margin of the Midland basin onto the shallower Eastern shelf. This pattern probably resulted from the gradual eastward dissipation of wave and current energy through absorption by bottom sediment. The main factor determining sediment texture appears to have been distance from unrestricted marine waters. All the facies present at Nena Lucia reflect shallow-water environments, probably no deeper than 100 ft. Lateral variation in one or more of the parameters affecting deposition or organic growth would produce unlike environments, in which discrete sedimentary and biotic patterns could develop in juxtaposition—in essence, facies. The present superposition of these facies is interpreted as the result of migration of environments through time.

STATISTICAL ANALYSIS OF PETROGRAPHIC DATA

To gain a better understanding of the facies, the petrographic data from all five wells were combined for statistical analysis aimed at (1) classifying each sample from all wells according to facies, (2) giving quantitative information on the composition of each facies, and (3) gaining information on the variations within facies. This statistical analysis showed that 82 percent of all samples studied can be classified in the facies assemblage outlined. The remaining 18 percent of the samples are gradational among facies.

Figure 4, prepared from the results of the statistical study, shows the average composition of

Table 1. Facies Differentiation Data

Parameter	Characteristics	How Estimated
Matrix	Grain or matrix supported Sparry Micritic	Estimated to nearest 10 percent
Grain type	Intraclasts Pelletoids Skeletal grains	Estimated to nearest 10 percent
Fossils	Platy algae Fusulinids Crinoidal debris	Estimate relative abundance (rare, common, abundant)
Secondary features	Recrystallization Porosity Dolomitization Oil stain	Rare, common, abundant Poor, fair, good, excel. Rare, common, abundant Light, medium, heavy
Impurities	Argillaceous material Quartz silt and sand Glauconite	Estimate relative abundance (rare, common, abundant)

the 23 parameters measured in the petrographic study of each facies. The average of at least one of these parameters is distinctly different for each of the facies.

Probably the most important result gained from the statistical study allows quantitative compositional limits to be set for the facies. From the statistical data, modified slightly by experience gained through the petrographic study, a system of classifying core samples can be proposed. The following limits have been used successfully in a study of core chips from many wells.

1. Crinoidal facies—more than 40 percent skeletal grains; crinoidal debris abundant (over 35 percent).

2. Pelletal-foraminiferal facies—more than 15 percent pellets, less than 25 percent intraclasts; crinoids and phylloid algae rare to common (less than 35 percent of each).

3. Algal plate facies—more than 40 percent skeletal grains; phylloid algae abundant (more than 35 percent).

4. Algal-intraclastic facies—more than 25 percent intraclasts, less than 15 percent pellets, less than 40 percent skeletal grains; phylloid algae common (5-35 percent).

5. Intraclastic facies—more than 35 percent intraclasts, less than 15 percent pellets, less than

30 percent skeletal grains; phylloid algae rare (less than 5 percent).

6. Micritic facies—more than 40 percent micrite.

SUMMARY OF PERTINENT METHODS AND CRITERIA FOR FACIES DIFFERENTIATION

The subdivision of the Desmoinesian section at Nena Lucia into facies can be made on a lithologic basis. However, the recognition of lateral variations within a facies, which may reflect direction or proximity to a buildup, appears to be dependent upon paleontologic criteria. The following criteria were found to have proximity implications and to be most usable for facies differentiation.

Lithologic Criteria

Although cores or core chips would be highly desirable, the important lithologic criteria can be observed satisfactorily in good cuttings. The only equipment necessary is a good microscope with a power range up to ×50. In addition, the samples should be cleaned with an acid rinse, and run wet. Table 1 shows the petrographic data that should be logged for facies differentiation. These tabulated data should be plotted adjacent to the lithologic log for comparison.

Table 2. Faunal Data

Fossil Types	Facies in which Important
A. Most Usable Megafossils	
Fenestellid bryozoans Fistuliporid bryozoans Prismoporid bryozoans	Crinoidal, algal-intraclast, and algal-plate
Inarticulate brachiopods Articulate brachiopods (many genera)	Crinoidal and algal-intraclast
Rugose and tabulate corals	Algal-intraclast and to some extent in crinoidal
B. Microfossils with Variable Lateral Distribution (from acid residues)	
Siliceous sponge spicules Conodonts	Crinoidal
Agglutinated foraminifers	Crinoidal and intraclastic
C. Microfossils with Variable Lateral Distribution (from thin sections)	
Calcareous mobile foraminifers	Crinoidal, algal-intraclast, and pelletal-foraminiferal
Calcareous encrusting foraminifers	Common in all; most characteristics in algal-plate
Fusulinids	In all facies; most common in pelletal-foraminiferal and crinoidal

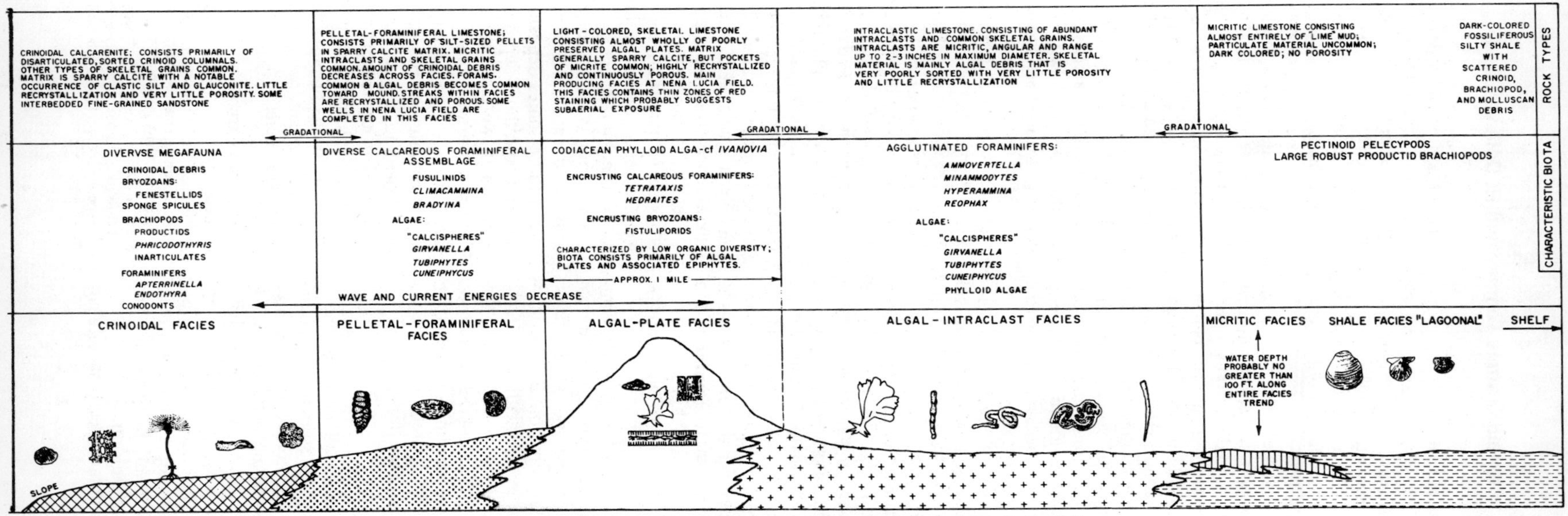

ROCK TYPES

CRINOIDAL CALCARENITE; CONSISTS PRIMARILY OF DISARTICULATED, SORTED CRINOID COLUMNALS. OTHER TYPES OF SKELETAL GRAINS COMMON. MATRIX IS SPARRY CALCITE WITH A NOTABLE OCCURRENCE OF CLASTIC SILT AND GLAUCONITE. LITTLE RECRYSTALLIZATION AND VERY LITTLE POROSITY. SOME INTERBEDDED FINE-GRAINED SANDSTONE

PELLETAL-FORAMINIFERAL LIMESTONE; CONSISTS PRIMARILY OF SILT-SIZED PELLETS IN SPARRY CALCITE MATRIX. MICRITIC INTRACLASTS AND SKELETAL GRAINS COMMON. AMOUNT OF CRINOIDAL DEBRIS DECREASES ACROSS FACIES. FORAMS COMMON & ALGAL DEBRIS BECOMES COMMON TOWARD MOUND. STREAMS WITHIN FACIES ARE RECRYSTALLIZED AND POROUS. SOME WELLS IN NENA LUCIA FIELD ARE COMPLETED IN THIS FACIES

LIGHT-COLORED, SKELETAL LIMESTONE CONSISTING ALMOST WHOLLY OF POORLY PRESERVED ALGAL PLATES. MATRIX GENERALLY SPARRY CALCITE, BUT POCKETS OF MICRITE COMMON. HIGHLY RECRYSTALLIZED AND CONTINUOUSLY POROUS. MAIN PRODUCING FACIES AT NENA LUCIA FIELD. THIS FACIES CONTAINS THIN ZONES OF RED STAINING WHICH PROBABLY SUGGESTS SUBAERIAL EXPOSURE

INTRACLASTIC LIMESTONE CONSISTING OF ABUNDANT INTRACLASTS AND COMMON SKELETAL GRAINS. INTRACLASTS ARE MICRITIC, ANGULAR AND RANGE UP TO 2-3 INCHES IN MAXIMUM DIAMETER. SKELETAL MATERIAL IS MAINLY ALGAL DEBRIS THAT IS VERY POORLY SORTED WITH VERY LITTLE POROSITY AND LITTLE RECRYSTALLIZATION

MICRITIC LIMESTONE CONSISTING ALMOST ENTIRELY OF LIME MUD; PARTICULATE MATERIAL UNCOMMON; DARK COLORED; NO POROSITY

DARK-COLORED FOSSILIFEROUS SILTY SHALE WITH SCATTERED CRINOID, BRACHIOPOD, AND MOLLUSCAN DEBRIS

CHARACTERISTIC BIOTA

DIVERSE MEGAFAUNA
CRINOIDAL DEBRIS
BRYOZOANS:
FENESTELLIDS
SPONGE SPICULES
BRACHIOPODS
PRODUCTIDS
PHRICODOTHYRIS
INARTICULATES
FORAMINIFERS
APTERRINELLA
ENDOTHYRA
CONODONTS

DIVERSE CALCAREOUS FORAMINIFERAL ASSEMBLAGE
FUSULINIDS
CLIMACAMMINA
BRADYINA
ALGAE:
"CALCISPHERES"
GIRVANELLA
TUBIPHYTES
CUNEIPHYCUS

CODIACEAN PHYLLOID ALGA-cf IVANOVIA
ENCRUSTING CALCAREOUS FORAMINIFERS:
TETRATAXIS
HEDRAITES
ENCRUSTING BRYOZOANS
FISTULIPORIDS
CHARACTERIZED BY LOW ORGANIC DIVERSITY; BIOTA CONSISTS PRIMARILY OF ALGAL PLATES AND ASSOCIATED EPIPHYTES.

AGGLUTINATED FORAMINIFERS:
AMMOVERTELLA
MINAMMODYTES
HYPERAMMINA
REOPHAX
ALGAE:
"CALCISPHERES"
GIRVANELLA
TUBIPHYTES
CUNEIPHYCUS
PHYLLOID ALGAE

PECTINOID PELECYPODS
LARGE ROBUST PRODUCTID BRACHIOPODS

CRINOIDAL FACIES | PELLETAL-FORAMINIFERAL FACIES | ALGAL-PLATE FACIES | ALGAL-INTRACLAST FACIES | MICRITIC FACIES | SHALE FACIES "LAGOONAL" | SHELF

WAVE AND CURRENT ENERGIES DECREASE

Approx. 1 mile

WATER DEPTH PROBABLY NO GREATER THAN 100 FT ALONG ENTIRE FACIES TREND

GRADATIONAL

SLOPE

Fig. 17—Idealized reconstruction showing facies relations across Nena Lucia field.

Paleontologic Criteria

As only a limited part of the fauna can be studied adequately in cuttings, cores or core chips are necessary if the full potential of the fossils is to be realized. If cores or core chips are available, the following paleontologic data should be recorded as indicated in Table 2.

The Foraminifera should be identified to the generic level and a distinction between mobile and encrusting types attempted. A 500-g sample can be crushed and dissolved in 10 percent formic acid for a period of 24 hours. The residue should be examined for conodonts, agglutinated Foraminifera, and siliceous sponge spicules. Thin sections (2 × 2 in.) may be prepared so that prismoporid bryozoans, fusulinids, and other calcareous foraminifers can be identified. Lateral changes in the faunas can be marked by the complete disappearance of taxa; however, in most cases, this change is reflected in total relative abundance. Therefore, total foraminiferal counts (total specimens on a thin section or from acid residues) are necessary.

If cuttings only are available, the use of paleontology is severely restricted. As many lateral changes are reflected only in variations in total relative abundance, contamination by caving is a serious problem. This is especially true in acidization, where 500-g samples should be used. Therefore, we believe that the only part of the entire fauna that can be studied reliably in cuttings is the assemblage of calcareous foraminifers and fusulinids. For these, composite thin sections of picked cuttings considered to be in place can be made.

The successful use of paleontology for proximity purposes in the middle and upper Strawn (Desmoinesian) requires some specialized skills, such as training in fossil recognition and familiarity with thin-section and formic acidization techniques. A paleontologic study as outlined in this report is dependent upon the availability of cores. Therefore, in areas where proximity information based on paleontology is desired, it is recommended that the appropriate section be cored.

Conclusions

Detailed studies based on all available information show that the upper Strawn (Desmoinesian) limestone section in the Nena Lucia area can be subdivided into meaningful facies.

Six different limestone facies have been delineated within the model area, of which four appear to have restricted distribution.

1. Crinoidal facies consists predominantly of crinoidal debris and is found in the outer and upper parts of the mound buildup.

2. Pelletal-foraminiferal facies consists of a significant content of pellets with some skeletal and intraclastic components and a conspicuous calcareous foraminiferal assemblage, is found on the immediate front (northwest) side of the carbonate bank, and contains porous streaks and is productive at places.

3. Algal-plate facies consists predominantly of phylloid-algal debris, evidently forms the massive porous core of the mound buildup, and is the main producing facies of Nena Lucia field.

4. Algal-intraclast facies contains abundant intraclasts, with a significant skeletal component consisting primarily of algal debris and agglutinated foraminifers; it is found in areas beneath and in back (southeast) of the mound buildup.

5. Intraclastic facies consists predominantly of micritic intraclasts with no appreciable skeletal content; it is interbedded with other facies.

6. Micrite facies consists primarily of carbonate mud, may be interbedded with other facies, but most typically is developed in the lower part of the algal-intraclast facies and the underlying Odom Formation.

All facies have environmental significance even though all were deposited in shallow water. All could have been, and possibly were, deposited simultaneously in juxtaposition through lateral variation in those parameters affecting sedimentation and organic growth. Figure 17 shows an idealized reconstruction of facies relations across the Nena Lucia field area and summarizes the characteristic rock types and biota.

Detailed studies of cores allowed us to define gross facies quantitatively. Once the core studies were completed, facies identification could be made from bit cuttings.

The mound buildups are enclosed in dark shale, and subtle differences can be noted between those shales deposited on the front side of the bank and those deposited on the back.

Although we can determine individual facies, the basic overall lateral lithologic patterns of the facies do not show enough variation to allow us more accurately to pin-point relative position on, or nearness to, a potential mound buildup. Data presented in this report have demonstrated that certain paleontologic components appear to show small variations of possible value in determining a drilling target, which in this instance is the algal-plate facies.

References Cited

Harrington, J. W., and E. L. Hazlewood, 1962, Comparison of Bahamian land forms with depositional topography of Nena Lucia dune-reef-knoll, Nolan County, Texas: Am. Assoc. Petroleum Geologists Bull., v. 46, p. 354-373.

Kerr, S. D., Jr., 1969, Algal-bearing carbonate reservoirs of Pennsylvanian age, west Texas and New Mexico (abs.): Am. Assoc. Petroleum Geologists Bull., v. 53, p. 726-727.

Jay Field, Florida—A Jurassic Stratigraphic Trap[1]

R. D. OTTMANN,[2] P. L. KEYES,[3] and M. A. ZIEGLER[4]

Abstract The first Jurassic oil discovery in Florida was made in June 1970, in Santa Rosa County near Jay, 35 mi (56.3 km) north of Pensacola. Current estimates indicate recoverable reserves in the Smackover Formation of 346 million STB of oil and 350 Bcf of gas. The accumulation occurs on the south plunge of a large subsurface anticline, and the updip trap is formed by a facies change from porous dolomite to dense micritic limestone.

The Smackover consists of a lower transgressive interval of laminated algal-mat and mud-flat deposits and an upper regressive section of hard-pellet grainstones. Early dolomitization and freshwater leaching have provided a complex, extensive, high-quality reservoir. Irregular distribution of facies presents difficult problems in development drilling, unitization, and pressure-maintenance programs.

Hydrogen sulfide content of the hydrocarbons requires expensive processing facilities and well investment. A typical completed well costs $650,000, and an additional $200,000 is required for flow-line and inlet separation facilities. Add to this $550,000 for plant facilities to sweeten the oil for market, and each well investment approaches $1,400,000.

Daily production from Jay field is 93,500 bbl from 89 wells. The rapid development of this field resulted from a drilling program coordinated with modular plant design.

INTRODUCTION

Jay field is the most significant discovery in the United States since the Prudhoe Bay field discovery of 1968. Current estimates indicate recoverable reserves of 346 million STB of oil and 350 Bcf of gas. Since discovery of Jay field in Santa Rosa County, Florida, 35 mi (56.3 km) north of Pensacola, industry has drilled more than 200 wildcats in southwestern Alabama and the Florida Panhandle. Most Jurassic traps in this area are very subtle, and the key to exploration lies in the ability to combine the prediction of occurrence of favorable facies with the delineation of low-relief structures. Without question, many significant oil and gas accumulations remain to be found in this area and in other areas with similar subtle relationships. This discovery has indeed proved that large undiscovered accumulations still exist in the United States.

HISTORY OF JURASSIC SMACKOVER EXPLORATION

Exploration of the Jurassic Smackover Formation can be divided into four major periods. The first includes the years 1937–1950, in which 34 fields, containing about 250 million bbl of oil in reservoirs between 6,000 and 12,000 ft (1,830 and 3,658 m), were discovered in southern Arkansas and northern Louisiana (Fig. 1). These fields occur in the updip part of the Jurassic embayment and produce from a widespread carbonate grainstone facies. Shallow depths and the effectiveness of conventional seismic methods provided the keys to this early period of successful exploration.

During the period from 1950 through 1960, attention shifted to the East Texas basin, where exploration met with only limited success because of restricted distribution of reservoir facies and the inability to define by seismic means deep structures in the central part of the basin.

Interest in Smackover exploration was rejuvenated from 1960 to 1968 as a result of the development of common-depth-point (CDP) seismic techniques, which extended structural definition to as deep as 25,000 ft (7,620 m), and the development of capability to drill to these depths. Dur-

[1]Manuscript received, October 17, 1974. Reprinted in revised form with permission from Gulf Coast Association of Geological Societies Trans., 1973, v. 23, p. 146-157.

[2]Exxon Co., U.S.A., Midland, Texas 79701

[3]Exxon Co., U.S.A., Houston, Texas 77001.

[4]American Arabian Oil Co., Dhahran, Saudi Arabia.

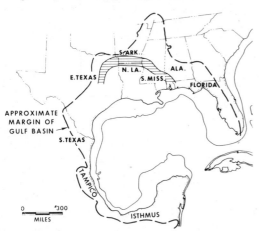

FIG. 1—Smackover producing trend, northern Gulf basin.

FIG. 2—Index map of Jay field area.

ing this period, exploration shifted to southern Mississippi, where 38 new fields were discovered with reservoirs at depths ranging from 12,000 to 18,000 ft (3,658 to 5,486 m).

Exploration for Jurassic Smackover accumulations in Alabama and Florida was a logical extension along trend of the discoveries of the first three periods. The fourth period began in 1968 with a Norphlet discovery on a well-defined structure at Flomaton in Escambia County, Ala-

bama (Fig. 2), 90 mi (145 km) east of the closest Jurassic production. Discovery of the Jay field followed in 1970 when dolomitized Smackover was found on a structural nose approximately 7 mi (11.3 km) south of Flomaton. Other significant discoveries during this period were Womack Hill, Chatom, Big Escambia Creek, Fanny's Church, and Chunchula fields in Alabama and the Blackjack Creek field in Florida.

REGIONAL STRATIGRAPHY

Figure 3 shows the Jurassic section penetrated in southern Alabama and the Florida Panhandle. The Louann Salt, of probable Middle Jurassic age, is overlain by Norphlet clastic rocks. Above the Norphlet is the Smackover, a carbonate section which is overlain in turn by the evaporites of the Buckner. On top of the Buckner is the upper Haynesville Formation, which consists of red, fine- to coarse-grained clastic units interbedded with evaporites. West of Jay the upper Haynesville Formation also contains carbonate rocks. This sequence is capped by the coarse-grained, in places gravelly, clastic beds of the Cotton Valley Group.

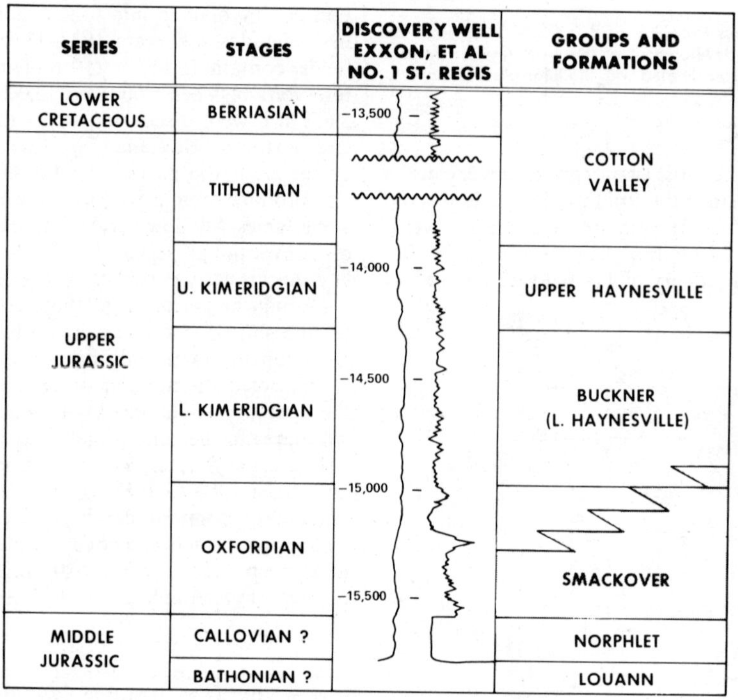

SERIES	STAGES	DISCOVERY WELL EXXON, ET AL NO. 1 ST. REGIS	GROUPS AND FORMATIONS
LOWER CRETACEOUS	BERRIASIAN	−13,500	
UPPER JURASSIC	TITHONIAN		COTTON VALLEY
	U. KIMERIDGIAN	−14,000	UPPER HAYNESVILLE
	L. KIMERIDGIAN	−14,500	BUCKNER (L. HAYNESVILLE)
	OXFORDIAN	−15,000 −15,500	SMACKOVER
MIDDLE JURASSIC	CALLOVIAN ?		NORPHLET
	BATHONIAN ?		LOUANN

FIG. 3—Stratigraphic nomenclature, Jay field, Alabama and Florida.

FIG. 4—Jay field, regional setting at end of Smackover deposition.

The Appalachian structural trend (Fig. 4) and other pre-Jurassic tectonic elements, such as the Conecuh and Pensacola ridges, had significant influence on the distribution of Jurassic sediments and were the source areas for basal Jurassic clastic material and for clastic sediments deposited during the "Smackover" transgression. Adjacent to the Appalachians, the Smackover is composed of a mixture of clastic and carbonate rocks. Major production in the Jurassic trend is obtained basinward from the area characterized by this mixed lithology, primarily in areas containing grainstone facies. The depositional environment of these grainstones is one of the focal points of this study.

As shown in Figure 5, the updip limits of the Louann Salt and the Smackover Formation reflect the topography of the pre-Jurassic surface. The shallow-water carbonate muds of the Smackover were deposited on a broad shelf in the area of the Florida Panhandle—a shelf which narrowed considerably toward Mississippi. Widespread distribution of shallow-water carbonate material on this shelf provided ideal conditions for the development of stratigraphic traps.

In Mississippi, the producing Smackover reservoir is generally an oolitic facies. In southern Alabama and in the Florida Panhandle, the grainstone facies is composed of hardened pellets; oolites constitute a very minor part of the section. Development of this low-energy facies in the Jay area can be attributed to the dissipation of wave energy over a wide shelf and to the absence of topographic features which would have created a focal point for the generation of oolites.

Figure 5 also shows the distribution of the Cotton Valley clastic beds, which overlap the pre-Jurassic surface for a considerable distance east of the Smackover updip pinchout.

COMPARISON OF JURASSIC DEPOSITION AND RECENT ENVIRONMENT

To increase understanding of the sedimentary processes prevailing during the Late Jurassic, and of the resulting depositional patterns, we have compared the Jay area with a modern carbonate province which it closely resembles—the Great Bahama Bank, particularly the Andros platform. Facies and depositional patterns here are similar to those in the area studied in the Florida Panhandle, although the climate during the closing phase of the Smackover sedimentary cycle was obviously more arid than the present climate of the Bahamas.

The sedimentary pattern of the Andros platform, shown in Figure 6, is based on descriptions of Holocene facies by Newell et al (1959) and Purdy (1963). The bank edges are shown to drop off abruptly into surrounding deeper water, and the bank margin is characterized by the presence of coralgal carbonate sand.

On the western side, extensive oolite shoals border the platform, spilling over more landward, protected, hard-pellet sediments. This sediment type represents a facies of carbonate sand composed predominantly of fecal pellets. The pellets

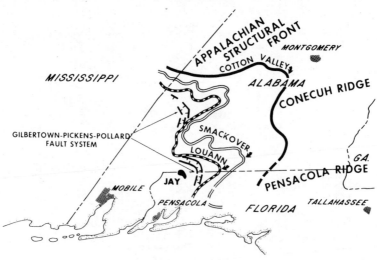

FIG. 5—Jay field, regional depositional limits.

are grains which became indurated either in a very shallow subtidal environment or during intermittent exposure at times of low tide on intensely burrowed stable flats. Such stable flats are distributed extensively along the protected interior side of all the bank islands.

The areas of grapestone with aggregate grains are located at the northern and southern ends of the platform. The area containing pelletoid carbonate muds is on the western or leeward side of Andros Island.

A good correlation exists between the facies distribution and varied sediment types of the Smackover at Jay and the stable-flat area west of Joulter Cays. The detailed facies distribution in this area, mapped by Purdy (1963) and others, is shown in Figure 7; the reef-covered bank margin is adjacent to the Tongue of the Ocean on the right and the northern tip of Andros Island at the lower left.

The reefs and the associated coralgal-sand-covered lagoon characterize the deeper marine environment of deposition on the east. In the Joulter Cays area, sediments accumulate in less than 3 ft (1 m) of water in the shallow shoals. The seaward side of these shoals is bordered by oolite bars and tidal deltas. These bars and tidal deltas protect the adjacent shallow stable-flat area, where grapestone bars, small emerging islands, and tidal flats are developed. Tidal channels are superimposed on these deposits.

The sediments of the shallow stable-flat areas are intensely burrowed by shrimplike decapods, which produce countless burrow mounds of digested and extruded sand-size pellets and skeletal fragments. The grass blades on such stabilized flats are inhabited by numerous species of Foraminifera, Bryozoa, and other organisms; algal balls or oncolites are also present. Identical lithologies with comparable fauna and flora have been observed in the lower part of the Smackover Formation.

SMACKOVER DEPOSITION

The postulated sedimentary conditions during deposition of the Smackover in the Jay area are shown in the block diagram in Figure 8. A coastal complex is shown with two very shallow embayments flanked by shallow subtidal slopes which outline the adjacent, burrowed stable flats. Above this sequence and closer to shore lie the algal mats of the closing phase of carbonate deposition. The arid supratidal flats represent the landward areas where evaporites were being formed while carbonate deposition was occurring farther seaward.

The basal Smackover carbonate rocks are characterized by a sequence of alternating laminites and pelletal-oncoidal limestone, grading upward to a more micritic pelletal unit. The laminites reflect a stromatolitically bedded sequence deposited in an intertidal to low supratidal environment. The burrowed pelletal-oncoidal deposits seem to correlate best with sediments deposited in modern grass-covered, stable-flat environments. Near the middle of the Smackover interval, a micritic

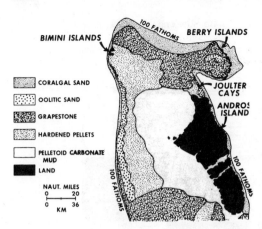

FIG. 6—Carbonate depositional pattern, Andros platform, Great Bahama Banks.

skeletal limestone marks the maximum extent of the marine transgression. The leached pelletal dolomite interval in the upper half of the Smackover represents deposition during the regressive cycle in an environment intermediate between shallow stable flats and more marine conditions. The uppermost Smackover laminites reflect renewed deposition in supratidal and intertidal environments.

The cuts of the block diagram depict the vertical and lateral distribution to the various sediment types. The patterns reflect a slow marine transgression during deposition of the lower Smackover. The regressive phase, which began after the deposition of the micritic skeletal limestone, is characterized by an infill of topographic lows and by shallow marine embayments.

Distinguishing features of the four major rock types making up the Smackover are illustrated in the photomicrographs of Figure 8:

1. Algal stromatolite, with laminated bedding;
2. Pelletal-oncoidal micrite, with the symmetry of an algal oncolite;
3. Fine skeletal micrite, with fossil hash;
4. Pelletal grainstone, shown with ovoid pellets of varied sizes.

SMACKOVER POROSITY

In Figure 9, the present porosity distribution is shown in relation to the depositional component parts of the Smackover. The same block diagram is used as that in Figure 8. Porosity formation by dolomitization is associated with the presence above the Smackover of the evaporite deposits of the Buckner Formation. Hypersaline waters, re-

sulting from isolation and evaporation of seawater along the shallow coastline, percolated through the underlying formation by flood recharge. Leaching by continental fresh waters followed as the coastline shifted seaward. In areas of grain carbonate rocks, the process was most effective in creating high-quality reservoir rock. In micritic and pelletal-oncoidal facies, some porosity was created, but areas of dolomitization and later leaching are more localized, mainly along the paths of fluid flow.

The three major porosity types and their relative locations are illustrated by the photomicrographs in Figure 9.

1. Grain moldic dolomite is the leached end product of a dolomitized hard-pellet grainstone.
2. Intercrystalline dolomite represents the leached product of a dolomitized micritic pelletal limestone (packstone to wackestone).
3. Leached matrix is the result of dolomitization of an algal stromatolite.

An example of the porosity distribution and its relation to the lithology of the Smackover Formation in the Jay field area can be seen in Figure 10, which is a reproduction of part of a density log from the Exxon-LL&E No. 1 Jones McDavid, the confirmation well for the field. The Smackover

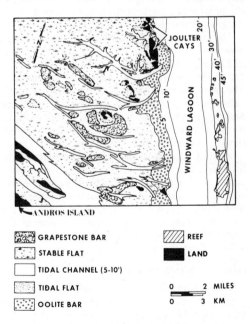

FIG. 7—Facies distribution, Joulter Cays area, Andros platform, Great Bahama Banks.

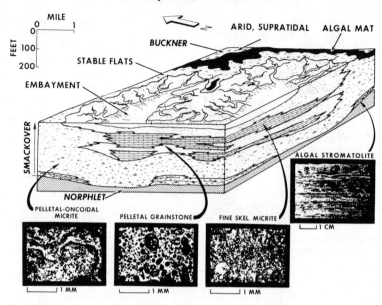

FIG. 8—Smackover facies, Jay field area.

has a thickness of slightly over 370 ft (113 m). The bottom contact of the Smackover Formation with the fine-grained Norphlet sandstone is typically sharp, whereas the transition at the top into the Buckner evaporites is gradational.

The facies distribution shown is similar to that presented in the block diagrams; the laminites and pelletal-oncoidal deposits are at the base, and the micritic skeletal limestone is in the middle. The main porosity is associated with the hard-pellet grainstone facies in the upper part of the Smackover. This zone of secondary moldic porosity was created by freshwater leaching in a superficially dolomitized pelletal limestone. Development of patchy, leached matrix porosity in the lower transgressive part of the Smackover is associated with possible intermittent surface exposure. The homogeneous character of the main Smackover reservoir rock is demonstrated by the smooth density curve. The lower portion of the reservoir, however, is broken by "tight" intervals, which reflect the occurrence of the laminites.

STRUCTURE AND TRAP

The accumulation at Jay field occurs on a south plunge of a large subsurface anticline, and the updip trap is formed by a facies change. The map of the structure at the top of the Smackover (Fig. 11) shows a combination structural-strati-graphic trap with 420 ft (128 m) of oil column extending over an area 7 mi (11.3 km) long and 3 mi (4.8 km) wide. There is only 100 ft of structural relief on the Jay nose above the saddle separating it from the large Flomaton structure adjacent on the north. The Flomaton anticline, which produces gas from the Norphlet, rises another 350 ft (107 m) to its crest. Other structurally significant features at Jay include the steep northeast flank, the Gilbertown-Pickens-Pollard fault on the east, and the small closures along the crest of the nose. The latter trap oil in the underlying Norphlet, but reserves are only about 1 percent of those of the Smackover.

Updip termination of favorable Smackover reservoir rocks occurs almost perpendicular to strike and reflects a boundary between depositional environments that remained constant for an extended period. North of the saddle, the rock is a dense micritic limestone; on the south, dolomitized grain carbonate rock is present. The productive area covers roughly 14,400 acres (58.4 km²).

Cross sections A–A' and B–B' (Fig. 12) show the irregular distribution of the porosity and its relation to present structure. The porosity build-ups, mainly in the hard-pellet facies, may be explained as resulting from the normal depositional process of filling the shallow embayments. Another explanation for the irregular thicknesses of

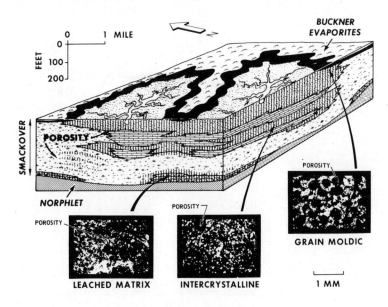

FIG. 9—Smackover porosity, Jay field area. Present porosity distribution is indicated by vertical hachures.

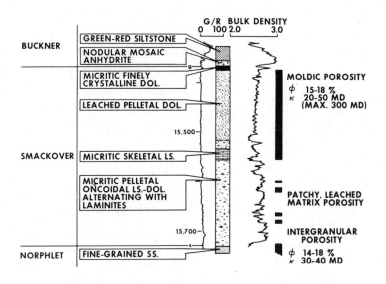

FIG. 10—Log of sedimentary sequence in Exxon-LL&E No. 1 Jones McDavid well, Jay field.

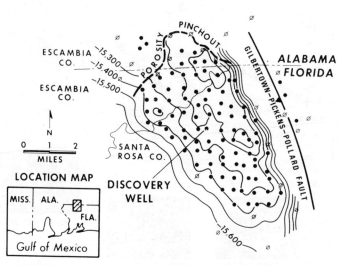

FIG. 11—Jay field, structure map of top of Smackover. C.I. = 100 ft.

porous rock would be the natural selection of environments by the organisms contributing to this facies and buildups resulting from extended stable conditions.

RESERVOIR PROPERTIES

An isopach map of the net feet of "pay" with greater than 8 percent porosity is shown in Figure 13. On the northwest end of the field, the productive limit corresponds to the porosity pinchout. Elsewhere, it follows an oil-water contact at −15,475 ft (−4,717 m). The contact varies in individual wells, which creates difficulties in totaling net "pay." As an example, the oil-water contact in two adjacent wells on the west flank—the LL&E No. 2–2 and No. 35–2 McDavid Lands — differed in elevation by 100 ft (30 m). It is possible that this irregular oil-water contact may be due to the limited volume of water associated with this reservoir. In an irregular development of porosity lenses, the presence of a limited quantity of water certainly would provide such a condition. Current reservoir data indicate a water volume of only one fourth of the oil volume.

On the north, the abruptness of the porosity pinchout was documented when the LL&E No. 33–3 McDavid Lands encountered only 9 ft (2.7 m) of "pay" in the original hole and 109 ft (32.2 m) in a horizontal sidetrack of 700 ft (213.4 m). A similar situation occurred on the east flank, where the Amerada Hess No. 40–16 Findley encoun-

tered no "pay" only 2,500 ft (761 m) from a well with 243 ft (74 m) of "pay."

The thickest net-"pay" interval is on the eastern, or shoreward, side of the field (Fig. 13); it thins abruptly toward the northeast and more gradually toward the southwest. It is apparent that the net-"pay" thickness corresponds to the distribution of the hardened pellets in the section. Subsequent dolomitization provided greater porosity and permeability on the east side, where hardened pellets are concentrated, than in any other area of the field. One well, the Exxon No. 5–2 St. Regis, drilled rock with permeability greater than 8,000 md. Average permeability for the reservoir is 35 md. Porosity ranges up to 31 percent and the average is 13 percent. Average "pay" thickness is 100 ft (30.5 m).

HYDROGEN SULFIDE

Crude oil in the Jay field contains 9 percent hydrogen sulfide, 3 percent carbon dioxide and nitrogen, and 88 percent hydrocarbons. Free sulfur, which has been found in impermeable parts of the Smackover, is possibly an alteration product of anhydrite. It is logical, therefore, to postulate that free sulfur was present in some of the sediments which subsequently were altered by dolomitization, and it was this free sulfur that furnished the source of the hydrogen sulfide in the crude. The free sulfur apparently was randomly distributed. Both south and east of Jay, sweet

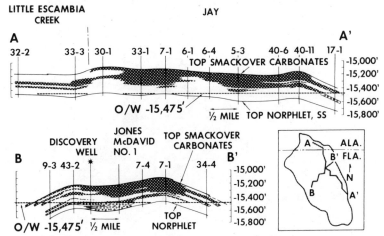

FIG. 12—Porosity-distribution cross sections of Little Escambia Creek and Jay fields.

crude is produced from Norphlet sandstones, which underlie the Smackover. In other characteristics, the crude is similar to that from the Smackover.

PRODUCTION

Production results from rock and fluid expansion. In general, the productivity of wells in this field will be restricted more by the 3½-in. (8.89 cm) tubing through which they produce than by reservoir quality. Original bottomhole pressure was 7,850 lb, or about 700 lb above saltwater gradient; saturation pressure was approximately 5,000 lb lower—2,830 lb. At reservoir conditions, the Jay crude is a mobile, low-viscosity liquid that will flow 2½ times more freely than formation water.

A comparison of the Jay Smackover reservoir with some well-known fields around the world is shown in Figure 14. Middle East field "A," productive from a Cretaceous dolomite, has reserves of approximately 3 billion bbl of oil. Middle East field "B" produces from lower Eocene carbonate rock and has reserves of more than 2.5 billion bbl of oil. Wasson field, in West Texas, produces from a Permian dolomite and has reserves of 1.5 billion bbl. The East Texas field, with original reserves of more than 5 billion bbl of recoverable oil, is included as a point of reference for a sandstone reservoir.

Of the fields compared, Jay shows the most rapid decrease in water saturation as height above free-water level is attained. In addition, a consistent relation of decreasing water saturation with increasing permeability exists at Jay. Even at 2 md permeability, the reservoir contains an irreducible water saturation of less than 15 percent.

This relation reflects less porosity volume occupied by connate water in the "pay" zone. It can be assumed that these low saturations are evidence of a clean reservoir (without typical water-retaining constituents) and will result in lower residual oil saturations at depletion.

These unique properties possibly result from a uniform rhombohedral porosity throughout the matrix which will contribute to a higher yield of oil per acre-foot. Field-wide unitization became effective March 1, 1974. Pressure-maintenance operations in the form of water injection will recover twice as much oil as will be produced from primary recovery.

DRILLING

Initially, drilling was difficult because of the Buckner salt and evaporite section overlying the Smackover, and because of the hydrogen sulfide in the hydrocarbons. Strict safeguards had to be employed at all times. As a result, a directional survey was run above the Smackover in each hole to provide a target in the event that the well blew out. To penetrate the Buckner required increasing the salt content of the mud to a level which would minimize solution but still provide a medium in which logs could be run.

Operating costs in the Jay area are high. Well costs range from $650,000 for a producing well to $375,000 for a dry hole. Flow-line costs vary with

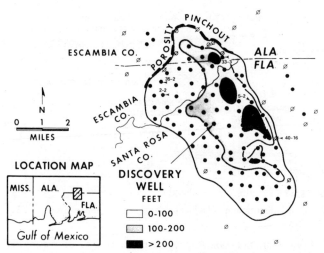

FIG. 13—Jay field, isopach map of Smackover net "pay" with porosity greater than 8 percent.

length, but an average cost with inlet separation facilities is $200,000. When each well's share of a 12,000 bbl/day processing facility costing $7 million is added, the investment per well reaches $1.4 million. Based on field rules established in Alabama and Florida providing for 160-acre spacing and allowables prior to unitization of 1,000 bbl/day per well, and considering current prices of $8.08/bbl for crude, 44.5¢/Mcf for 1,350-BTU gas, and $13/LT for sulfur, revenue equaled investment for a producing well in slightly less than 5 months.

To date, 89 producing wells and 18 dry holes have been drilled in the Jay field. Major operators include Exxon, Louisiana Land and Exploration, Amerada Hess, Sun, Chevron, and Moncrief and Young.

Total field investment to date, including the dry holes and processing facilities, but not the exploration and leasing costs, amounts to approximately $132 million.

PROCESSING

The flow through a sweetening facility similar to those at Jay is shown schematically in Figure 15. Each full well stream enters two separators in series, from which the oil is fed to an oil-stabilization unit where heat is added and the hydrogen sulfide is vaporized from the crude. The sweetened crude oil then goes to the stock tank. The

gas from the two separators and the oil-stabilization unit separately enters three gas-treating towers where diethanolamine, a liquid with a natural affinity for hydrogen sulfide, absorbs the hydrogen sulfide from the gas. Sweet gas leaving the gas-treating towers goes to the dehydrator and then to the gas market. Sour gas stripped from the diethanolamine is carried to a Claus sulfur-recovery unit, where 96 percent of the sulfur is recovered by heating and cooling the acid gas

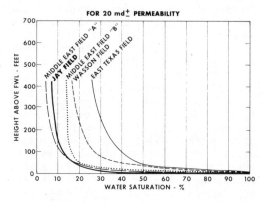

FIG. 14—Reservoir comparison of Jay field and several other fields. Capillary pressure, plotted as height above free-water level, is compared with permeabilities of about 20 md and porosities of 15-20 percent.

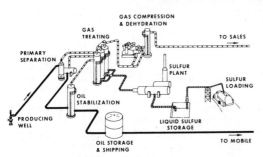

FIG. 15—Schematic flow diagram of production facilities at Jay field.

through several critical temperature ranges. After treatment, the hydrogen sulfide levels are essentially zero. A 12,000-bbl plant will produce, in addition to the oil, 12 MMcf of sales gas and 80 long tons of sulfur per day.

The treated crude is transported through a 16-in. (40.6 cm) line from Jay to Mobile. The major part of the gas is sold to Florida Gas Transmission Company. Sulfur is sold to Freeport Sulphur Company and is trucked in liquid state to Mobile. Cumulative production from the Jay field as of June 1, 1974, was 58 million bbl of oil and 70 Bcf of gas.

One of the key decisions in developing a field that requires processing facilities is determining the proper size of plant to be built. The specific sizes constructed result from the location and size of the operator's leases, the economy of the operation, and the pressure of competitive production. Six 12,000-bbl and three 6,500-bbl plants have been built at Jay, providing a total of 91,500 bbl of daily processing capacity. Because of the extra capacity normally designed in such facilities, the processing facilities at Jay can handle the production, as permitted under unitization, of 93,500 bbl of oil per day. The gross investment for the nine plants is approximately $50 million.

ENVIRONMENTAL MONITORING

Prior to the beginning of production, a baseline environmental survey was conducted in the area surrounding the plants. A year later a second study was made to determine any changes in the environment. At the same time, a system was set up to determine routinely the air quality. Currently, 10 air-monitoring stations are established throughout the area to report the sulfur dioxide

and hydrogen sulfide concentrations to the state on a weekly basis.

CONCLUSION

Development of Jay field has required a level of expertise above that commonly required. The timeliness of most critical decisions was made possible as a result of early field definition. Step-out locations defined the approximate productive limits within 18 months of discovery. As a result, it was possible to start construction of the plants before many of the development wells were drilled. These early decisions and the modular plant design have allowed this field to come on stream in an extremely short period of time.

The results of this venture have led to other significant discoveries in the southern Alabama–Florida Panhandle area, and it is probable that the geologic understanding gained from study of Jay field will have application in other areas.

SELECTED REFERENCES

Bathurst, R. G. C., 1969, Bimini Lagoon, in H. G. Multer, Field guide to some carbonate rock environments, Florida Keys and western Bahamas: Miami Geol. Soc., Rosenstiel School of Marine and Atmos. Sci., Univ. Miami, Miami, Florida, p. 62-69.

Keyes, P. L., 1971a, Jurassic geology of Flomaton field area of southern Alabama (abs.): AAPG Bull., v. 55, no. 2, p. 347.

———1971b, Geology of the Jurassic, Flomaton-Jay area, Alabama and Florida (abs.): Gulf Coast Assoc. Geol. Socs. Trans., v. 21, p. 30.

Kinsman, D. J. J., 1969, Modes of formation, sedimentary association, and diagnostic features of shallow water and supratidal evaporites: AAPG Bull., v. 53, no. 4, p. 830-840.

Logan, B. W., et al, 1970, Carbonate sedimentation and environments, Shark Bay, Western Australia: AAPG Mem. 13, 223 p.

Newell, N. D., et al, 1959, Organism communities and bottom facies, Great Bahama Bank: Am. Mus. Nat. History Bull., v. 117, art. 4, p. 179-228.

Ottmann, R. D., P. L. Keyes, and M. A. Ziegler, 1973, Jay field—a Jurassic stratigraphic trap (abs.): AAPG Bull., v. 57, no. 4, p. 748.

Purdy, E. G., 1963, Recent calcium carbonate facies of the Great Bahama Bank—1: Jour. Geology, v. 71, p. 334-335; pt. 2: Jour. Geology, v. 71, p. 472-497.

Traverse, A., and R. N. Ginsburg, 1966, Palynology of the surface sediments of Great Bahama Bank, as related to water movement and sedimentation: Marine Geology, v. 4, no. 6, p. 417-459.

Geology of Fairway Field, East Texas[1]

ROBERT T. TERRIERE[2]

Abstract The Fairway field, Anderson and Henderson Counties, Texas, is a major oil field in a reef and reef-associated facies of the Lower Cretaceous James Limestone Member of the Pearsall Formation. The present productive limits are controlled largely by structure. The location of the reef also was influenced by a contemporaneously growing structure, so the reservoir can be considered to have a combination structural-stratigraphic trap, both physically and genetically.

The James Limestone Member consists of several limestone types, differentiated on the basis of texture and fossil content in cores. Maps of the distribution of these rock types during successive stages of reef growth show that the main core of the reef, dominated by frame-building organisms, was in the northwest part of the field. The frame-builders were a closely associated suite of corals, stromatoporoids, algae, and rudistids. By about the middle of the time of development of the James reef, the center of growth was at its maximum, and smaller satellite reefs appeared in the southeast and southwest. A facies dominated by large bivalves occupied much of the area between centers of growth of the main frame-builders. The south-central part of the field was an area of persistent accumulation of carbonate sands and gravels. Carbonate muds and muddy sands were the dominant facies elsewhere.

Porosity and permeability are present in all of the limestone types but are higher on the average in the associated limestones than in the reef proper. The porosity is largely secondary, although it is in part the result of enlargement of primary pores.

INTRODUCTION

The Fairway field is located in Anderson and Henderson Counties, Texas, approximately on the axis of the East Texas basin (Fig. 1). The principal oil and gas production is from the James Limestone Member of the Pearsall Formation, although smaller amounts have been produced from the Ferry Lake Anhydrite and the Rodessa and Sligo Formations. All of these units are of Early Cretaceous age (Fig. 2). Only the James is included in the present study.

The reservoir rock in the James Limestone Member is an unusually good example of a subsurface reef complex. Despite widespread recognition of the importance of reefs in petroleum exploration, relatively few limestone bodies are sufficiently cored or exposed—or have original textures well enough preserved—to demonstrate convincingly a reef origin. Material for the present study included cores from about 35 wells in various parts of the Fairway field. Electric logs

from about 200 wells were used to determine the overall shape of the limestone body and the structure and thickness of various units.

The field was discovered in 1960. By the end of 1974 it had produced 123,703,000 bbl of oil and had estimated reserves of 76,234,000 bbl (Oil and Gas Jour., 1975, p. 118). The oil is 48° API gravity. Other reservoir properties and production practices are cited by Perkins (1964) and by Calhoun and Hurford (1970).

In the immediate area of the Fairway field are two much smaller fields. For this report, the Frankston field, just east of Fairway, is considered a part of the Fairway field; however, the reservoirs probably are separated by a fault. The Isaac Lindsey field, just southwest of Fairway, is an unimportant gas field producing from the James limestone.

The trap in the Fairway field is a combination structural and stratigraphic trap. The location of the reef complex in the James was at least partly determined by a growing structure. Later vertical movements modified the shape and extent of the trap.

The major stratigraphic units in the vicinity of the field are shown in Figure 2. The Hosston Formation, which overlies the generally accepted Jurassic-Cretaceous boundary, is a terrigenous unit that contains redbeds and conglomerate, especially in the lower part. Above the Hosston are three limestone units—the Sligo, James, and Rodessa—separated by intervals of dark gray calcareous shale. The limestone units range from very dark and argillaceous to lighter colored and relatively pure. The purer facies of the Sligo and Rodessa are largely grainstones, some of them oolitic and others composed of fossil fragments, oncolites, and Foraminifera. The James is regionally extensive as a thin argillaceous limestone. Where the

[1]Manuscript received, February 27, 1975. Published with permission of Cities Service Oil Company.

[2]Cities Service Oil Company, Tulsa, Oklahoma 74110.

The writer thanks the many colleagues who assisted or supported various phases of the study, especially T. L. Broin, M. K. Horn, N. P. Leiker, W. K. Pooser, B. A. Silver, and K. F. Wantland, Jr.

reef facies is well developed in the Fairway field, the James is as thick as 250 ft (75 m) and is relatively pure and porous.

Overlying the Rodessa Formation is the Ferry Lake Anhydrite, which consists of anhydrite with interbeds of shale and limestone. The limestone is largely argillaceous, but isolated thin beds arc oolitic and contain intraclasts, pellets, small mollusks, and Foraminifera. These limestone beds apparently are the reservoirs for the small amount of oil that has been produced from the Ferry Lake, although cementation and partial replacement by anhydrite have reduced their porosity and permeability.

JAMES LIMESTONE MEMBER

Structure and Thickness

Structure of the top of the James limestone reef is shown in Figure 3. The basic configuration is that of a southeast-plunging nose, truncated on the northwest by a group of faults and somewhat modified by other faults on the east and northeast.

The thickness of the James limestone, including argillaceous limestone at the base, is shown on Figure 4. The thickness generally parallels the structure. The overall parallelism of isopach lines and contours of the top of the reef appears to be the result of localization of reef growth by the developing structure. The thickness differences are much less than the elevation differences, so the shape of the top of the limestone is not simply depositional. The conclusion that the structure was growing during the time of deposition is also supported by isopach maps of thin arbitrary intervals above and below the reef. These maps show patterns similar to those of the reef thickness and structure, especially with regard to considerable thickening northwest of the Fairway field. The shales and shaly limestones in these intervals surely were deposited in horizontal layers, so the thickness variations must reflect active structural movement during deposition.

Salt domes are present in the general area of the field (see Fig. 1), and it is probable that the persistent vertical movements were caused by

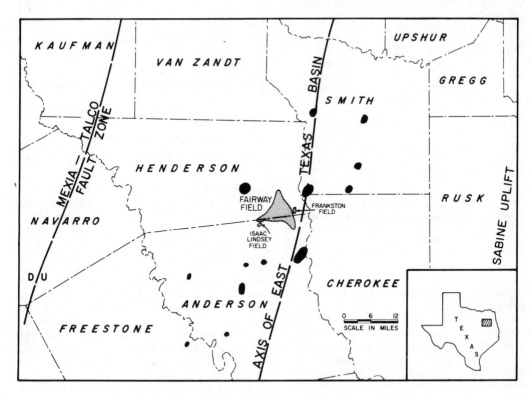

FIG. 1—Location of Fairway field, Texas. Black spots are salt domes.

flowage of the underlying salt into the domes. It has been suggested by geologists working in the area that the Fairway field is on an "interdomal high" still underlain by the original thick salt after flowage of salt from adjacent areas into the domes. Others believe that the area is a "turtle-shaped structure" (Trusheim, 1960, p. 1533) underlain by relatively thick post-salt deposits that had filled a topographic low during early salt flowage.

Lithologic Varieties

For purposes of showing the major rock types in the field, a somewhat informal limestone classification was established based on megascopic description of slabbed cores. During description of the first few cores, it was seen that certain varieties of limestone kept recurring. Each major variety was defined as a limestone type and identified by a number. Description of additional cores resulted in some modification and additions to the system. The present scheme seems to work well in expressing the major limestone types in the Fairway field and in showing vertical and lateral changes.

The rock types recognized are as follows:

Type I.   Calcarenite and calcirudite
    a.   Conglomerate and well-sorted carbonate sandstone (grainstone of Dunham, 1962)
    b.   Muddy carbonate sandstone (packstone)
Type II.   Predominantly micritic limestone, mostly with intermixed carbonate sandstone and shell fragments (wackestone and mudstone)
Type III.   Limestone containing large bivalves
    a.   Many unbroken shells, commonly upright in the rock
    b.   Broken shells, some algal-coated
Type IV.   Limestone characterized by colonial organisms such as corals, stromatoporoids, rudistids, and algae
Type V.   Dark argillaceous limestone

Type I

The limestones included in Type I are the carbonate sandstones and conglomerates—the definitely clastic limestones (see Figs. 5, 6). The coarsest grains are mostly intraclasts (terminology of Folk, 1962). They apparently were not transported far, although many are well rounded. Nearly all seem to have been lithified when transported; none is squashed as would be the case if reworked while soft. Among the pebble-sized fossils, clam

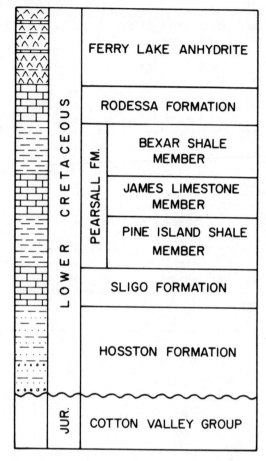

FIG. 2—Partial columnar section, Fairway field area.

fragments appear to be the most abundant, but other types of fossils seen in the reef are also present, as are many unidentifiable fragments. Sand-size grains are more difficult to identify. They include intraclasts, pellets, and fossils; the relative abundance of each is uncertain because much of their character has been obliterated by micritization. Many are the intermediate-type grains, neither intraclasts nor pellets, that have been called "lumps" or "pelletoids" (called "peloids" by McKee and Gutschick, 1969, p. 101, 555). The interstitial material includes both micrite and spar cement. Many limestones have microspar cement, which appears to be mostly primary rather than recrystallized micrite.

Type I limestones can be further subdivided into Types Ia and Ib. The calcirudites and rela-

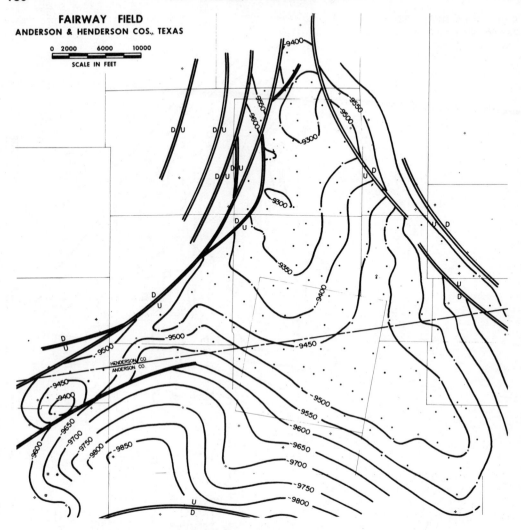

FAIRWAY FIELD
ANDERSON & HENDERSON COS., TEXAS

0   2000   6000   10000
SCALE IN FEET

FIG. 3—Structure of top of James Limestone Member. Depths are in feet below sea level.

tively coarse calcarenites are characterized by spar or microspar cement and rounded grains. They constitute the limestone Type Ia and were deposited under conditions of considerable wave action, probably as marine bars.

The limestones classed as Type Ib are transitional into the calcareous mudstones of Type II. They have an appreciable proportion of micrite matrix and are generally poorly sorted and not well rounded. Environmentally, they also are intermediate between Types Ia and II. For convenience in description, rare intervals of alternating Types Ia and II were also classed as Type Ib.

Type II

Type II limestones are composed predominantly of micrite (Figs. 7, 8). They include the wackestone and mudstone of Dunham's (1962) terminology and the calcilutite of older usage. Some rocks in Types III, IV, and V are also more than 50 percent micrite, so Type II is a "wastebasket" group for micritic limestones that do not fit into one of the other categories. Moreover, some rocks that megascopically appear to be micrite proved to contain many pellets and other grains when studied in thin section.

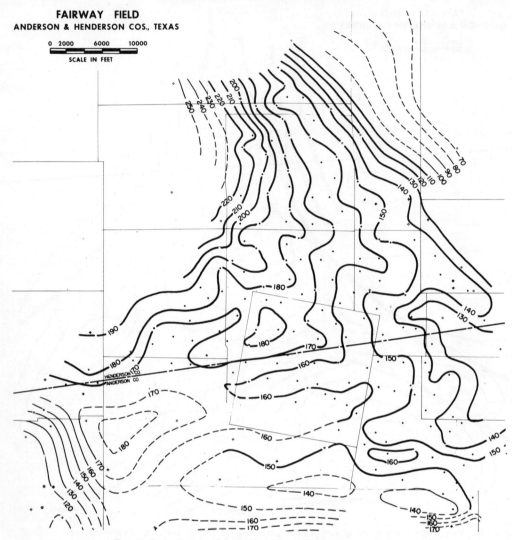

FAIRWAY FIELD
ANDERSON & HENDERSON COS., TEXAS

0 2000   6000   10000
SCALE IN FEET

FIG. 4—Isopach map of James limestone. Thicknesses are in feet.

Environmentally, most Type II limestone represents quiet-water deposition. It may have been deposited partly in sheltered intra-reef locations and partly in deeper water.

Type III

Some limestones in the Fairway area are distinctive because of their content of very large bivalves (Figs. 9-11). These constitute limestone Type III. Some of the clams appear to be about 6 in. (15 cm) across, although the limited size of the core prevents viewing of whole specimens. They have not been formally identified, but many re-

semble the genus *Chondrodonta*. Perhaps a variety of biologic types is grouped together, but smaller shells are specifically excluded. Some of the large clams are in upright, presumably growth position in the rock (Fig. 9). In other intervals, shells that are considerably broken seem to be mostly from the same type of very large clam (Figs. 10, 11).

Type III limestones are subdivided into Types IIIa and IIIb. Rocks of Type IIIa contain many relatively unbroken shells, many of which are upright. Type IIIb designates rocks containing broken shells. Many fragments are impossible to identify even as to the size of the original shell, so

FIG. 5—Type Ia limestone, a well-sorted grainstone (calcarenite). Black sootlike "gilsonite" in pores emphasizes rock texture. Core slab from Cities Service No. 1 Emerson, 9,876 ft (3,010 m).

greater biologic diversity is represented in Type IIIb than in IIIa. In some Type IIIb limestones the clam shells are riddled with holes bored by worms or other organisms. These shell fragments also are highly corroded, and some are coated with algae.

In Type IIIa the material between the clams typically is smooth-appearing micrite. Only rarely does it contain much carbonate sand or many fossils. The matrix of Type IIIb limestones contains sand-size carbonate grains, but carbonate mud predominates. Although Type IIIb is transitional into Type I limestone, borderline cases are surprisingly uncommon.

Some Type III limestones are transitional into Type IV. They contain some colonial corals but very few other frame-building organisms such as stromatoporoids. Type III limestone also borders on Type IV limestone in an environmental sense. The large clams that characterize it lived around the edges of the reef core and spread over the surrounding sea bottom, possibly at very shallow depths.

Type IV

All limestones characterized by frame-building organisms are grouped togehter in Type IV (Figs. 12, 13). This limestone forms the "true reef" or "reef-core" lithology that is capable of growing at angles steeper than the angle of repose and of forming wave-resistant structures because of the framework effect of colonial organisms. Several groups of colonial organisms are represented in the James limestone, including corals, stromatoporoids, algae, rudistids, and a few bryozoans. Type IV limestones could be subdivided further on the basis of the type of colonial organism present. Achauer (1967) and Achauer and Johnson (1969) showed a vertical progression of reef faunas from a stromatolite-hydrozoan zone (containing algal banding and the hydrozoans here referred to as "stromatoporoids") to a *Chondro-*

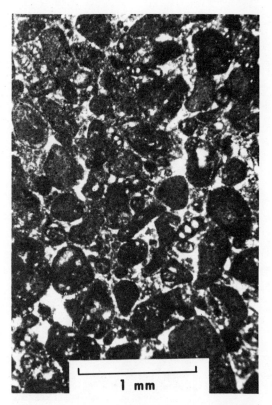

1 mm

FIG. 6—Type Ia limestone in thin section. Grains are pellets, intraclasts, and Foraminifera. Unidentifiable micritized fossil fragments are common in this lithology. From Hunt No. 1 West Poynor, 9,925 ft (3,025 m).

donta zone, and then to a rudistid zone. For the present study, however, the limestones containing frame-building organisms have all been grouped under Type IV.

Type V

Type V includes the limestones that are dark colored and impure (Figs. 14, 15). It is gradational into other varieties of limestone and into shale, and dividing lines are hard to place. For cores in which a differentiation of Type V limestone from calcareous shale was difficult, the decision was based on fissility. Rocks with a conchoidal or irregular fracture across the bedding were classified as Type V, and those that split along bedding planes were classified as shale.

The separation of Type V limestone from other limestone types was based largely on color, which also corresponds well with the amount of insolu-

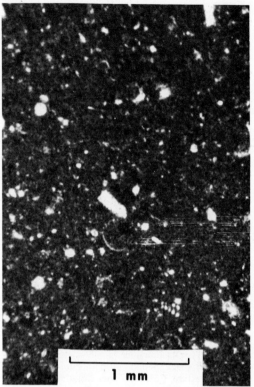

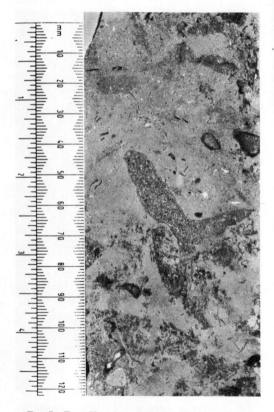

Fig. 7—Type II carbonate mudstone containing burrows filled with muddy carbonate sandstone. Core slab from Atlantic No. 1 Larue, 9,999 ft (3,048 m).

Fig. 8—Thin section of Type II carbonate mudstone containing scattered small fossils. Many such thin sections contain considerable sand-size material and suggestions of pellet texture. From Cities Service No. B-1 Miller, 9,851 ft (3,003 m).

ble material. Limestones of Type V are medium dark gray or darker, based on comparison of a freshly broken surface with the rock-color chart. Texturally, Type V limestones most commonly grade into the carbonate mudstones of Type II, but samples intermediate between Type V and Type Ib (muddy sandstones) are also common. Thin sections show that sand-size carbonate grains are more abundant than would be suspected from viewing hand specimens. Some Type IV limestones are fairly dark colored, but little difficulty was encountered in separating them from Type V.

Core-slab surfaces of Type V limestone that have been etched with hydrochloric acid have a dirty brownish-gray color, apparently caused by a surface residue of clay and by slight oxidation. Many of them show irregular sedimentary struc-

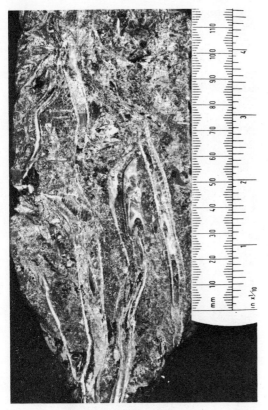

FIG. 9—Type IIIa limestone containing large bivalve shells upright in rock. Core slab from Cities Service No. C-1 Miller, 9,901 ft (3,018 m).

tures, probably from slight slumping of the sediment while soft, from compaction, and from burrowing organisms. Others have spheroidal to irregular concretionlike masses that are slightly more calcareous than the rest of the rock. Some of the Type V limestone contains a few megafossils, mostly small clams. Pyrite is common, and quartz silt is more abundant than in the cleaner limestone types. Some very dolomitic Type V samples from above and below the reef resemble the other Type V rocks.

Type V includes essentially all of the offreef James limestone. This type has not been divided into subvarieties, but recognition of meaningful variations in these rocks can be important in reef exploration. For example, a core of dark argillaceous limestone from a dry hole less than 1 mi (1.6 km) northeast of Fairway field contains sand-size, relatively pure intraclasts in Type V limestone at the stratigraphic level of the reef (Figs. 16, 17). The well was drilled before the reef was discovered, and in retrospect the intraclasts can be recognized as evidence of a nearby build-up, although none are identifiable as definite reef fragments.

Other Limestone Varieties

Some relatively minor limestone types are distinctive enough to merit special mention, though they are not differentiated as major rock types. At one time a limestone type was established for biomicrite containing many megafossils, separate from the types containing large clams (Type III) or frame-builders (Type IV). These fossiliferous limestones contain moderate to small-sized bivalves, gastropods, echinoderm fragments, and unidentifiable shell fragments, all of which are widespread in the James. Experience has shown

FIG. 10—Type IIIb limestone containing many broken bivalve shells and a few small corals. Core slab from Cities Service No. C-1 Miller, 9,891 ft (3,015 m).

that they are abundant enough to form a distinctive rock type in only a very few places, and the subdivision was dropped from the classification.

Some of the cores contain many oncolites, mostly pebble-sized pisolites composed of wavy algal bands enclosing rounded shell fragments or intraclasts (Fig. 18). Oncolites are numerous enough in some rocks to form a distinct limestone variety. The oncolite-rich lithology tends to overlap Type IIIb limestone by virtue of the presence of progressively thicker algal coatings on broken clam fragments. The volume of oncolite limestone is small, so oncolites were noted as a subsidiary constituent in rocks classified as one of the other types, usually IIIb. Environmentally, the oncolites seem to have formed in areas of relatively slow deposition, as indicated by their association with bored and corroded clam fragments;

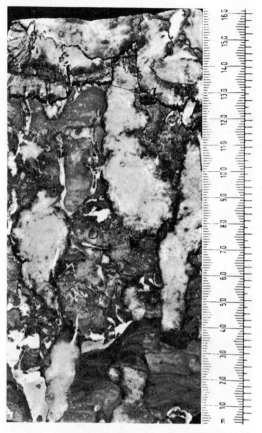

FIG. 12—Fingerlike or plumelike stromatoporoid colonies in Type IV limestone, "core" facies of reef. Dark bands of probable algal origin are locally visible in mud between colonies. Core slab from Cities Service No. C-1 Ellis, 9,936 ft (3,028 m).

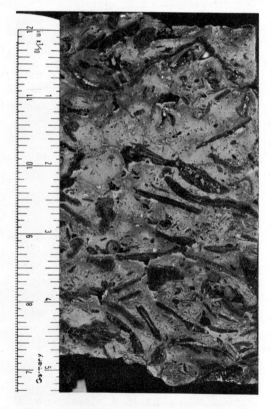

FIG. 11—Type IIIb limestone in which fragments of large mollusks are corroded and bored and heavily coated with algae. Core slab from Cities Service No. C-1 Miller, 9,886 ft (3,013 m).

it is supposed that algae were better able to coat grains that were not buried too rapidly.

Limestones rich in Foraminifera have not been classified separately. Foraminifera are sufficiently abundant in some cores to form distinctive rock types, but these are conveniently treated as varieties of Type I and Type II limestones, and the abundance of Foraminifera is noted separately. There is a remarkable relative scarcity of foraminifers in rocks of Type III and Type IV. Miliolids are the most abundant group of foraminifers in and near the reef, but other types are present also, especially in offreef areas. The Foraminifera are concentrated in the upper part of the James. It is largely speculative whether the environment in which they flourished was specifically provided

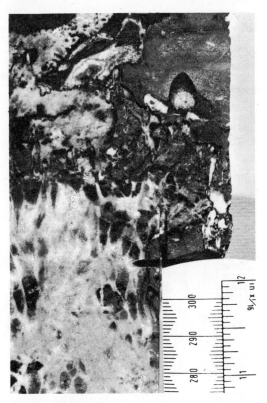

FIG. 13—Type IV limestone containing a largely re-crystallized coral colony. Algae and a small rudistid(?) also are present. Atlantic No. 1 Truitt, 10,264 ft (3,128 m).

by the reef itself, or whether some essentially independent ecologic factor such as regional water depth or temperature was favorable only during the latter part of reef growth.

DISTRIBUTION OF LIMESTONE TYPES

The three-dimensional distribution of the rock types is most conveniently shown by a series of maps, each indicating the lithologic distribution during a different vertical interval of reef growth. Ideally, the separation of intervals should be by time planes correlated from well to well. Unfortunately, no internal marker horizons can be recognized, and artificial "time" intervals had to be used. The engineering committee concerned with oil recovery in the field used sonic logs to correlate porosity zones between wells (Fairway Technical Comm., 1969) and mentioned geologic correlations by Achauer. Such correlations can be

made locally but become questionable at best over wider areas and are doubtful markers of contemporaneity.

Above and below the reef, many thin electric-log units are easily recognized across the field. The number and remarkable continuity of these units, which are composed of calcareous shale and argillaceous limestone, are convincing evidence that the lithologies above and below the reef parallel time planes. Subdivision of the reef limestone itself was done by dividing the interval between the highest marker below the reef and the top of the reef into six equal units (Fig. 19). These were treated as approximately contemporaneous units in the various wells, a working hypothesis that helps in visualizing the succession of facies patterns. The lowest one sixth of the interval between the markers is below the reef proper and has not been cored; it was labeled "unit 0" and was not used further. The other divisions, from the base up, were called "units 1 to 5." Their thickness varies from well to well in proportion to variation in the thickness of the entire James interval.

FIG. 14—Type V argillaceous limestone. Acid etching of smoothed core slabs produces a surface coating of clay much lighter colored than natural rock and brings out structures not otherwise easily seen. This core slab shows considerable burrowing; from Cities Service No. C-1 Miller, 9,825 ft (2,995 m).

The lithology in each "time" unit was determined from cores, and the dominant lithology was used where several limestone types were interbedded. The lateral boundaries shown between facies on the map were influenced by the abundance of subsidiary lithologies in mixed intervals. Electric logs were used to aid in refining lateral changes between shale and limestone in unit 1. The resulting maps are shown as Figures 20–24.

Unit 1

The distribution of the limestone types in the first of these units, representing initial stages of reef development, is shown on Figure 20. The apparent center for growth of frame-building organisms was in the northwest part of the present Fairway field, where a small area of Type IV limestone was deposited. The shape and extent of this reef framework are uncertain because of limited core control, but basic patterns of limestone-shale distribution are apparent from electric logs. Subordinate amounts of limestone with frame-builders in wells outside the shaded area also help in fixing approximate limits.

The same reasoning, plus somewhat better core control, places centers of calcarenite distribution

FIG. 16—Broken core surface that has been etched with dilute hydrochloric acid. Rock is Type V limestone, dark colored and very clayey, and etching has left a light gray insoluble clay residue standing out on surface. Intraclasts of purer limestone have dissolved somewhat, leaving smooth surfaces with no clay residue. Intraclasts could be seen on natural rock surface only with difficulty. From Cities Service No. 1 Xenia Miller, 10,138 ft (3,090 m).

in the south-central and southern parts of the field. The northern reef-core facies and southern calcareous sandstone facies are separated by an area of carbonate and terrigenous mudstone.

Unit 2

The second of the subdivisions of the James limestone reef is characterized by a greatly expanded reef core in the northwest and north-central part of the area and by the presence of relatively pure limestone over areas where argillaceous micrite had been deposited previously (Fig. 21). Calcarenites of limestone Types Ia and Ib are notably widespread, and the principal area of Type Ia deposition lies along an east-west barlike feature.

FIG. 15—Core slab, more typical than that in Figure 14, of dark argillaceous nonreef facies of James limestone (Type V). From Cities Service No. B-1 Miller, 10,047 ft (3,062 m).

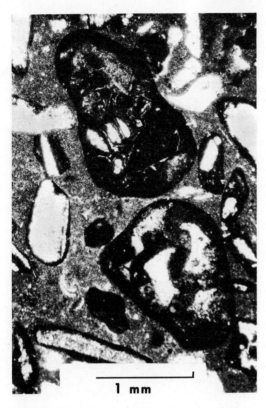

FIG. 17—Thin section of Type V limestone (see Fig. 16). Limestone intraclasts are in matrix of micrite and clay. Plane-polarized light. From Cities Service No. 1 Xenia Miller, 10,137 ft (3,089 m).

Unit 3

Limestones characterized by large fossils reached their maximum extent during Unit 3 deposition (Fig. 22). The main reef core of Type IV limestone in the northwest actually diminished slightly in size, but new centers of reef-core facies appeared in the southwest and southeast parts of the field. In addition, Type III limestone extended over a broad region, and large bivalves dominated the fauna between the centers of growth of corals, stromatoporoids, etc. Cross sections through the field show that this Type III limestone formed a definite unit that is traceable over several square miles. The ecologic significance of this extensive area of growth of large clams is uncertain, but its lateral extent would suggest that the seafloor was fairly flat. Perhaps the reef had grown nearly to sea level, which caused a pause in upward growth and a spreading out over a larger area. The common occurrence of borings and algal coatings in Type IIIb limestones may indicate relatively slow deposition. Flattening of the reef top and perhaps restriction in the interior would result in shift of growth of frame-builders to the margins of the reef. Extensive Type Ia limestone along the southern edge of the field indicates persistent wave action in that area.

Unit 4

During unit 4 deposition, growth of large organisms was somewhat more restricted and was largely confined to a band along the northeast side. The main reef core in the northern part of the field had shifted slightly to the northeast from earlier positions. Comparison of Figures 20 and 23 will show that areas that were offreef clayey limestones to the northeast of the reef in unit 1 were within the center of growth by the time of unit 4 deposition. Satellite centers of growth persisted, especially one on the east side of the field. On the southwest tip of the present field, an area of satellite reef growth in unit 3 was supplanted by Type III limestone in unit 4.

Another striking change during unit 4 deposition was the sudden influx of large numbers of Foraminifera, principally miliolids. The presence of numerous Foraminifera is indicated by an "F" on the maps. The foraminifers did not invade the centers of growth of the frame-builders but were abundant in the sheltered, and perhaps slightly restricted, area in the reef-complex interior. They are very numerous also in the sandstones in the southern part of the field, where they may have been concentrated partly by waves and currents.

Oncolites, the pisolite-like algal grains, are also relatively abundant in this interval. Their distribution is not shown on Figure 23, but it generally parallels that of the abundant Foraminifera.

Unit 5

The final stage of the reef is marked by further restriction of the reef core and the continued presence of abundant foraminifers and oncolites (Fig. 24). The northern center of growth of the frame-builders had almost disappeared, and its remaining remnant was farther northeast than before, overlying older offreef sediments. The satellite reefs on the east and southwest persisted in small areas. As indicated by Achauer (1967), rudistids largely supplanted stromatoporoids as frame-builders in this upper part of the reef complex. In the center of the field, micrites were deposited in greater abundance than before, and calcarenites were less common.

Porosity and Permeability

The porosity in the Fairway field is largely secondary, although primary porosity and permeability have influenced distribution of the secondary pores. In some samples, porosity is due chiefly to dissolution of grains, whereas in others matrix porosity is dominant. In still others, porosity is nonselective regarding fabric, occurring in both grains and matrix.

To compare the reservoir qualities of the various types of limestone, a tabulation was made of rock type versus porosity and permeability for 13 cores from which both lithologic descriptions and core analyses were readily available. The averages from this tabulation are shown in Table 1.

It is emphasized that there are many sources of error in this procedure, and the values should be considered approximate. They include values both from good reservoir rock and from denser zones within the reef. Rocks of Type V, the dark, argillaceous nonreef limestones, were not included because cores of obviously nonporous rocks are rarely analyzed in the core laboratories. This same bias in sampling also tends to increase the apparent porosity and permeability values of other relatively nonporous limestone types, because values that would be very low were not obtained. Permeability values have very great sample-to-sample variation, so the averages of permeability are less reliable than those for porosity. Other uncertainties arise from the fact that the analyses were run by three different laboratories using somewhat different techniques and methods of reporting.

Despite these problems, the figures can be used as a rough measure of the relative reservoir properties of the different types of limestone. Each of the rock types contributes to the reservoir. This is worth emphasis because of the widespread impression that reefs are far better reservoirs than are other carbonate rocks, and because of the tendency to equate the word "reef" with rocks containing frame-building organisms.

The highest average porosity and permeability values are for Type Ia and Type IIIb. Type Ia, the well-sorted calcarenite and calcirudite, has had considerable reduction of original intergranular porosity by cementation, but enough has remained to permit relatively easy development of secondary porosity. Type IIIb contains many fragments of mollusks, which probably were partly aragonite. The fragmental nature of the rocks and the mineralogic instability of the shells would both favor development of secondary porosity.

Fig. 18—Limestone composed of pisolites that apparently are oncolites formed by algal coating of shells. This lithology is transitional into algal-coated mollusk facies of Figure 11. From Cities Service No. 1 Miller, 9,863 ft (3,006 m).

Conversely, Type II limestones have relatively low porosity and permeability, probably because of low primary permeability. Type Ib limestone is intermediate between Type Ia and Type II, both in lithology and in porosity and permeability. In Type IIIa, large bivalves are relatively unbroken and commonly are in a micrite matrix; thus, it is intermediate in lithology between Types IIIb and II, as is indicated also by porosity.

Type IV limestones, those with frame-building organisms, are relatively low in porosity. The organisms are in a matrix of micrite, and many are surrounded and partly coated by dark bands of algal origin (Achauer and Johnson, 1969). The frame-building organisms themselves, especially the stromatoporoids, do not seem to have been easily leached, and their occurrence in relatively thick beds may also have minimized leaching.

As mentioned, the nonreef limestones of Type V have rarely been analyzed, but there is no

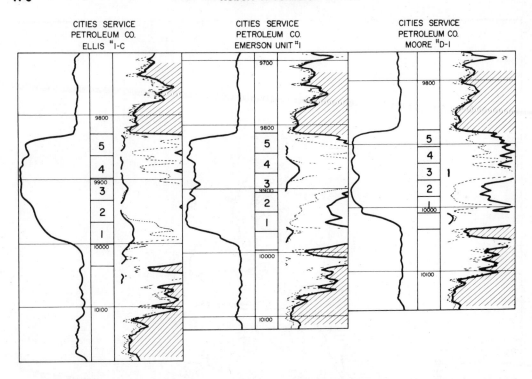

FIG. 19—Electric logs of James limestone interval from three representative Fairway field wells in northwest, central, and southeast parts of field. An arbitrary space under resistivity curve has been shaded to emphasize detail with which pre-James and post-James sections can be correlated. Within James limestone, subdivisions used for Figures 20-24 are numbered 1 through 5.

doubt that their porosity and permeability are very low.

ADDITIONAL FEATURES OF REEF LIMESTONE

Accessory minerals are a relatively minor component of the James limestone. The mineral dolomite is relatively scarce in the reef, and none of the rock is dolomite. The mineral occurs in two forms—as coarse white dolomite within the reef complex itself and as a very finely crystalline variety that is largely restricted to the argillaceous nonreef rocks. The coarse dolomite is mostly a replacement and cavity fill in megafossils. Petrographic evidence indicates an early postdepositional origin. Some of the limestones that contain this coarse dolomite also have small amounts of anhydrite.

The finely crystalline dolomite is most abundant in argillaceous pre-reef and post-reef beds and rarely exceeds a few percent of the rock volume. The impermeability of the rocks would suggest an early origin for this fine dolomite, but no direct evidence is available.

Chert is virtually absent from the James Limestone Member. Trace amounts of authigenic silica occur as tiny quartz crystals, most of them with nuclei of detrital quartz silt. The detrital quartz itself is largely confined to the argillaceous nonreef rocks; it is present in the reef only locally and in minor amounts.

Many parts of the James contain a black organic material, much of it a powdery "soot." Elsewhere, the material is less dispersed and is present in shiny masses with conchoidal fracture. Apparently both types are forms of the same material; they have been called "gilsonite" in some core and sample descriptions. They occur within secondary pores and hence have entered the rock after its original deposition. It seems probable that the original hydrocarbon to enter the pores has further matured into the present high-gravity oil and the sootlike black residue.

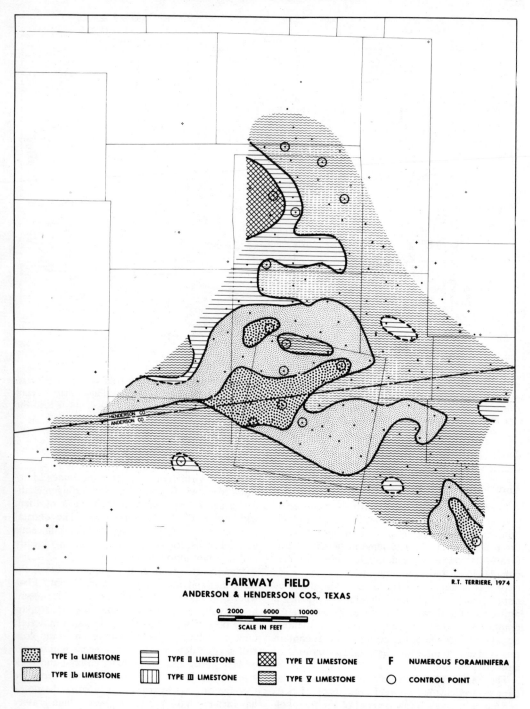

FAIRWAY FIELD
ANDERSON & HENDERSON COS., TEXAS

R.T. TERRIERE, 1974

0 2000 6000 10000

SCALE IN FEET

TYPE Ia LIMESTONE   TYPE II LIMESTONE   TYPE IV LIMESTONE   F NUMEROUS FORAMINIFERA

TYPE Ib LIMESTONE   TYPE III LIMESTONE   TYPE V LIMESTONE   ○ CONTROL POINT

FIG. 20—Distribution of limestone types in unit 1.

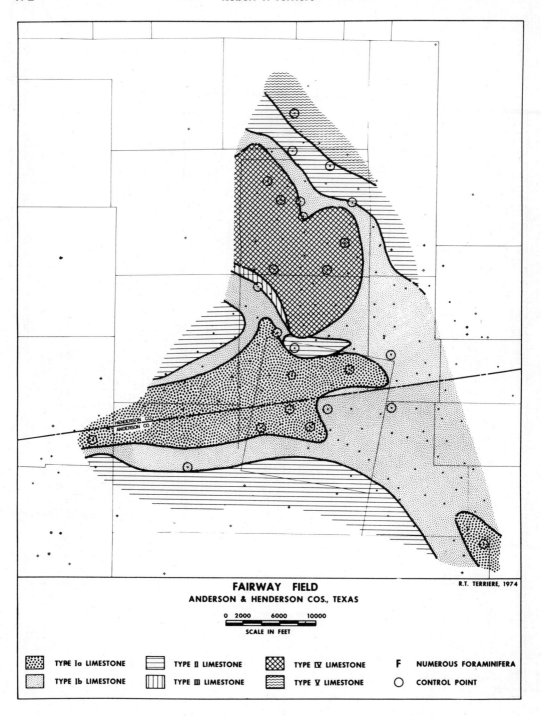

FAIRWAY FIELD
ANDERSON & HENDERSON COS., TEXAS

R.T. TERRIERE, 1974

0 2000   6000   10000
SCALE IN FEET

TYPE Ia LIMESTONE   TYPE II LIMESTONE   TYPE IV LIMESTONE   **F** NUMEROUS FORAMINIFERA

TYPE Ib LIMESTONE   TYPE III LIMESTONE   TYPE V LIMESTONE   ◯ CONTROL POINT

FIG. 21—Distribution of limestone types in unit 2.

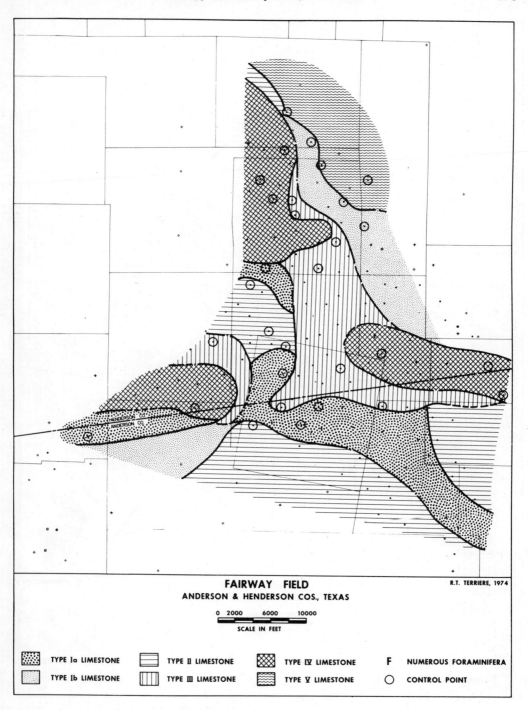

FAIRWAY FIELD
ANDERSON & HENDERSON COS., TEXAS

R.T. TERRIERE, 1974

0   2000   6000   10000
SCALE IN FEET

TYPE Ia LIMESTONE		TYPE II LIMESTONE		TYPE IV LIMESTONE	**F** NUMEROUS FORAMINIFERA
TYPE Ib LIMESTONE		TYPE III LIMESTONE		TYPE V LIMESTONE	◯ CONTROL POINT

FIG. 22—Distribution of limestone types in unit 3.

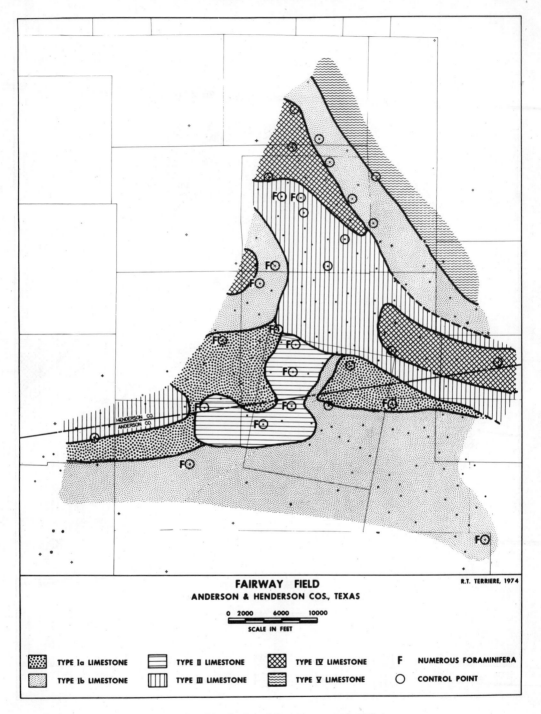

FIG. 23—Distribution of limestone types in unit 4.

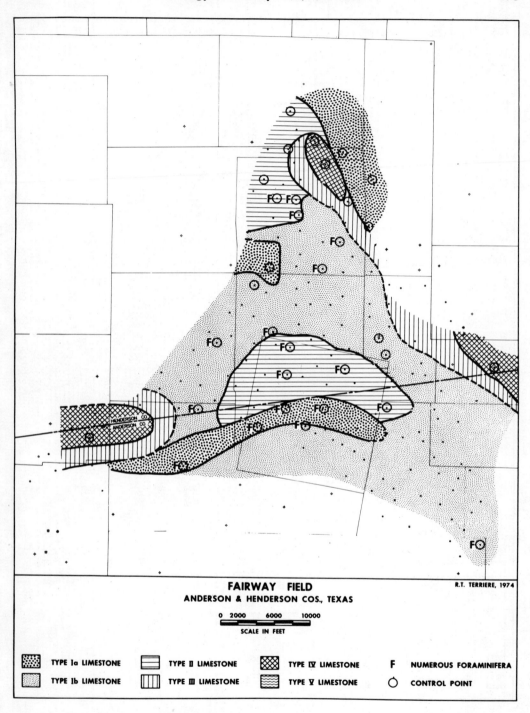

FAIRWAY FIELD
ANDERSON & HENDERSON COS., TEXAS

R.T. TERRIERE, 1974

0   2000   6000   10000
SCALE IN FEET

TYPE Ia LIMESTONE	TYPE II LIMESTONE	TYPE IV LIMESTONE	**F** NUMEROUS FORAMINIFERA
TYPE Ib LIMESTONE	TYPE III LIMESTONE	TYPE V LIMESTONE	CONTROL POINT

FIG. 24—Distribution of limestone types in unit 5.

Table 1. Average Porosity and Permeability Values, Fairway Field

Limestone Type	No. of Values	Average Porosity (%)	Average Permeability (md)
Ia	276	10.8	37.2
Ib	322	9.4	14.0
II	185	7.2	8.0
IIIa	87	9.2	28.6
IIIb	84	10.8	43.7
IV	106	8.2	12.4

CONCLUSIONS

The reef complex that forms the reservoir rock for the Fairway field grew on a slight topographic rise over a continuously positive structure. Later, more pronounced vertical movement resulted in faulting that restricts the present field area to only a portion of the reef complex.

The earliest deposition of the reef proper was at the northwest corner of the present field. This center of growth expanded and moved eastward, and smaller centers of reef growth appeared to the southeast and southwest. The south-central part of the field area was the site of persistent accumulation of carbonate sands and pebbles. Muddy carbonate sands and muds extended over broad areas around and between the other facies. During maximum development of the reef proper, the large interior area between centers of growth was inhabited by large bivalves. Near the end of reef growth, miliolid Foraminifera and oncolites were abundant in this interior area.

All of the limestone facies in and associated with the reef have porosity and permeability, and all contribute to the oil production. The produc-tive limits are defined mostly by faults and structural elevation. Only on the northeast edge of the field is there a facies change to nonreef argillaceous limestone.

Because of the sparsity of drilling outside the productive area, the areal limits of the reef limestone are not well known. The thickest James reef limestone lies west and north of the Fairway field proper. Perhaps additional oil is present in structural traps in other parts of this same reef complex.

REFERENCES CITED

Achauer, C. W., 1967, Petrography of a reef complex in Lower Cretaceous James limestone (abs.): AAPG Bull., v. 51, p. 452.
———— and J. H. Johnson, 1969, Algal stromatolites in the James reef complex (Lower Cretaceous), Fairway field, Texas: Jour. Sed. Petrology, v. 29, p. 1446-1472.
Calhoun, T. G., and G. T. Hurford, 1970, Case history of radioactive tracers and techniques in Fairway field: Jour. Petroleum Technology, v. 22, p. 1217-1224.
Dunham, R. J., 1962, Classification of carbonate rocks according to depositional texture, in W. E. Ham, ed., Classification of carbonate rocks: AAPG Mem. 1, p. 108-121.
Fairway Technical Committee, 1969, Fairway (James lime) unit stratification study: Unpub. rept. to operators in Fairway field.
Folk, R. L., 1962, Spectral subdivisions of limestone types, in W. E. Ham, ed., Classification of carbonate rocks: AAPG Mem. 1, p. 62-84.
McKee, E. D., and R. C. Gutschick, 1969, History of the Redwall Limestone of northern Arizona: Geol. Soc. America Mem. 114, 726 p.
Oil and Gas Journal, 1975, Here are the big U.S. reserves: Oil and Gas Jour., v. 73, no. 4, p. 116-118.
Perkins, S. L., 1964, Fairway field, in Occurrence of oil and gas in northeast Texas: East Texas Geol. Soc. Pub. 5, v. 1, p. 13-25.
Trusheim, F., 1960, Mechanism of salt migration in northern Germany: AAPG Bull., v. 44, p. 1519-1540.

Sunoco-Felda Field, Hendry and Collier Counties, Florida[1]

A. N. TYLER and **W. L. ERWIN**[2]

Abstract Sunoco-Felda field is located on the South Florida shelf, on the northeastern flank of the South Florida embayment. Production is principally from a stratigraphically trapped oil accumulation in a reefoidal, algal-plate, gastropod-bearing limestone mound in the Sunniland Limestone of Early Cretaceous age. The discovery well was drilled in July 1964 by Sun Oil Company on the basis of a combination of regional subsurface geology and geophysical work. The oil reservoir is about 11,475 ft (3,500 m) below the surface and is characterized by excellent vuggy porosity ranging upward to 28 percent; maximum permeability reaches 665 md. The field has a 34-ft (10 m) oil column and encompasses a surface area of approximately 4,500 acres (18 km²). In-place oil reserves are estimated to be 44 million bbl. The South Florida shelf area is sparsely drilled and offers great potential for the discovery of additional fields the size of Sunoco-Felda field. The subtle expression of this type of low-relief feature in the subsurface requires the complete coordination and application of sophisticated geological and geophysical techniques in order to provide a successful and economically attractive exploration program.

INTRODUCTION

Sunoco-Felda field is located in the southern part of the Florida Peninsula in Hendry and Collier Counties, about 25 mi (40 km) east of Fort Myers (Fig. 1). The oil accumulation is principally stratigraphic; it is in a reefoidal, algal-plate, gastropod-bearing limestone mound in the Sunniland Limestone of Early Cretaceous age. The top of the reservoir is at approximately 11,475 ft (3,500 m). Sun Oil Company drilled the discovery well in July 1964 on the basis of a combination of subsurface geology and seismic data. At the time of the discovery, only four wells had been drilled in Hendry County to the Sunniland Limestone in the search for oil. Indeed, the whole state of Florida had only one producing oil field within its borders—Sunniland field, about 15 mi (24 km) southeast of Sunoco-Felda, which produces from the Sunniland and was discovered in December 1943 by Humble Oil & Refining Company (now Exxon, U.S.A.). Forty Mile Bend field had been found in Dade County in 1954 but was abandoned in 1955. It, too, produced from the Sunniland Limestone.

The Felda prospect had been an area of interest to oil explorationists since the drilling in June 1954 of the Commonwealth No. 3 Red Cattle, in Sec. 25, T45S, R28E, Hendry County. This wildcat well, drilled with financial help from Sun Oil Company and Humble Oil & Refining Company, was located on a seismic anomaly. It recovered 1,090 ft (332 m) of oil and 9,060 ft (2,761 m) of salt water on a drill-stem test in the Sunniland Limestone. Since all information indicated this well to be marginal at best, a completion was not attempted and the hole was plugged and abandoned. After two additional attempts to discover a commercial oil field in the area, interest subsided and activity ceased. No further drilling took place in the area until 10 years later, when Sun Oil Company reassembled and enlarged its block of acreage and drilled the discovery well. Later, during the development of the field, a commercial oil well was completed by Sun approximately 1,000 ft (300 m) southwest of the abandoned Commonwealth No. 3 Red Cattle well.

PHYSIOGRAPHY

The surface is a featureless sandy plain with a maximum relief of only 10 ft (3 m) in the field proper. The higher elevations are generally pine- and palmetto-covered terraces of Pleistocene-Holocene sandstones and sands; the lower areas are marshy ponds and sloughs. Average surface elevation is about 32 ft (10 m) above sea level. There is no evident surface expression of the underlying oil-bearing feature. Surface drainage is northwest to the Caloosahatchee River. The area is directly north of the Big Cypress Swamp and just west of the Everglades.

Agriculture is the predominant economic activity of the area; beef cattle and citrus fruit are the principal year-round products. During the fall and winter months the vegetable-growing industry predominates. Much of the cultivated land is eventually turned into improved pasture providing lush and nutritious forage for a rapidly growing cattle industry.

[1]Manuscript received, January 29, 1975.

[2]Sun Oil Company, Dallas, Texas 75230.

Acknowledgment is given to Sun Oil Company for permission to publish this paper. The writers gratefully express appreciation to the many Sun Oil Company people who advised and assisted in its preparation, and especially to John A. Means for his technical assistance and helpful suggestions.

146

GENERAL GEOLOGY

The field is situated on the South Florida shelf, on the northeastern flank of the South Florida embayment, southwestward from the Peninsular arch (Fig. 2). Regional dip on the shelf proper at the Sunniland level is to the southwest at the rate of about 20 ft/mi (3.7 m/km). In specific areas where low-relief depositional structures or patch reefs are present, the local dip may reach 60–80 ft/mi (11–15 m/km). The entire shelf area, however, is characterized by uniform gentle dip with little or no evidence of structural deformation.

Depositional conditions across the shelf area appear to have undergone little change since Early Cretaceous time. This environment was one of shallow, clear, subtropical seas covering, in gener-

al, a slowly subsiding sea bottom. Lithologic uniformity of the resultant thick section of carbonate rock is evidence that the rate of deposition was approximately equal to the rate of subsidence. Local transgressions and regressions of the sea did occur, however, as evidenced by the presence of cyclic depositional sequences within the section and the existing types of lithology, which range from marine limestones to a dolomite-evaporite facies.

The Sunniland oil reservoir at Sunoco-Felda field appears to be a localized reef buildup of a part of a regional carbonate bank. Production is from an algal-plate, foraminiferal, pelletal limestone—a stacked biostromal unit. A quiet-water environment existed at the time of deposition. Tidal channels or passes probably cut the north-

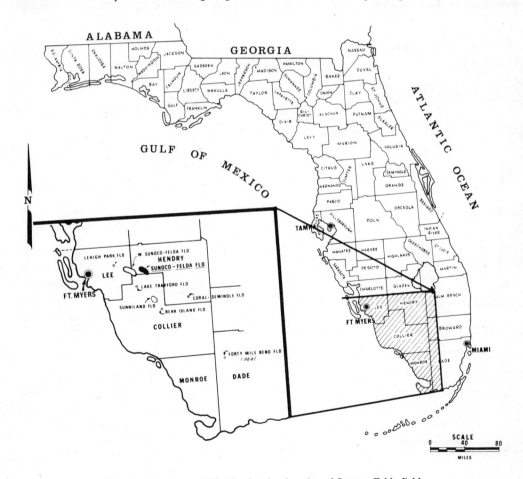

FIG. 1—Index map of Florida showing location of Sunoco-Felda field.

west-southeast-trending carbonate bank, separating the local reef pods. These channels eventually filled with carbonate muds, resulting in a loss of permeability and effectively separating one porous algal mound from another (Fig. 3).

Sunoco-Felda field is composed of two of these individual reef pods, both of which have a generally north-south trend. The most easterly pod contains by far the most porous and best preserved reservoir rock and is the significant oil-producing portion of the field. The Commonwealth No. 3 Red Cattle well, which tested the first oil in the area, is located on the western reef pod.

STRATIGRAPHY

The stratigraphic section below the surface Pleistocene-Holocene sandstones and sands consists predominantly of carbonate rocks and evaporites ranging in age from Early Cretaceous to Holocene. Figure 4 is a composite log of the sedimentary section.

Basement rock, although not penetrated in Sunoco-Felda field, was present at a depth of 15,670 ft (4,776 m) in the Humble No. 1 Lehigh Acres Development Corporation well, in Sec. 14, T45S, R27E, Lee County. This well is essentially along strike with, and approximately 8 mi (13 km) west-northwest of, Sunoco-Felda field. The basement complex found in this well—a gabbro-type mafic igneous rock—and the overlying 3,000 ft (915 m) of Lower Cretaceous carbonate rocks and evaporites are assumed to be similar to the section underlying Sunoco-Felda field. The deepest well in the field, Sunoco-Felda Unit No. 30-1, in the NE ¼, Sec. 30, T45S, R29E, Hendry County, was

FIG. 2—Map of Florida Peninsula showing major structural features (after Applin and Applin, 1965).

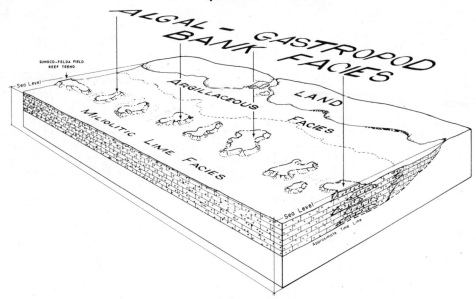

Fig. 3—Schematic drawing showing postulated shelf conditions during time of deposition of Sunniland, Sunoco-Felda reef trend.

in Lower Cretaceous limestone at the total depth of 12,686 ft (3,866 m). Of particular significance, near the bottom of this well, is the "Brown Dolomite," which occurs between 12,410 and 12,620 ft (3,783 and 3,847 m). This is a buff to brown, crystalline porous dolomite with interbedded limestone which could make an excellent oil reservoir.

Above the "Brown Dolomite" is the Punta Gorda Anhydrite of Trinity age. This is a 500-ft (152 m) massive gray anhydrite interbedded with irregular laminations of argillaceous limestone. It underlies the 50–60-ft (15–18 m) dark brown, highly fractured, dense, lithographic limestone unit which is the lowest member of the Sunniland. This is the oil-productive zone at the one-well Lake Trafford field in Collier County, about 9 mi (14 km) southwest of Sunoco-Felda field. Because it is highly fractured, this zone is usually recovered in cores as broken and shattered pieces of hard, brown limestone resembling a pile of rubble. As a result, the name "Rubble Zone" has come into common usage for this unit.

Above the "Rubble Zone" is a 30–40-ft (9–12 m) bed of black, very calcareous shale and/or carbonate mudstone. This shale is possibly the oil source for the overlying oil reservoir of Sunoco-Felda field.

Between the "Black Shale" and the "Upper Massive Anhydrite" lies a 150±-ft (45± m) sec-

tion of Sunniland Limestone (Fig. 5). In Sunoco-Felda field this is the oil-producing interval known as the "Roberts Zone." The lower portion of the "Roberts" is a tan to light gray, medium-hard, miliolid-bearing, chalky limestone which generally is slightly porous and has poor permeability. It is on this chalky limestone that the highly porous and permeable, bioclastic patch reefs developed; these patch reefs form the oil reservoir at Sunoco-Felda field. The producing interval is composed of gastropods, algal plates, assorted bioclastic debris, and limestone pellets (Figs. 6–9). These porous reef patches, or pods, grade both laterally and updip into miliolid-bearing carbonate mudstones with poor permeability. This facies change forms a permeability barrier which provides the updip limit of oil accumulation in the field.

The oil reservoir is capped by an impermeable brown limestone, 9–11 ft (2.7–3.3 m) thick, which in the local area effectively separates the producing zone from an overlying porous, saltwater-bearing, rudistid limestone. The caprock and rudistid limestone merge into one porous and permeable zone in West Sunoco–Felda field, 6 mi (9.6 km) to the west, and form an integral portion of the oil reservoir of that field.

The "Upper Massive Anhydrite," which is approximately 100 ft (30 m) thick, overlies the

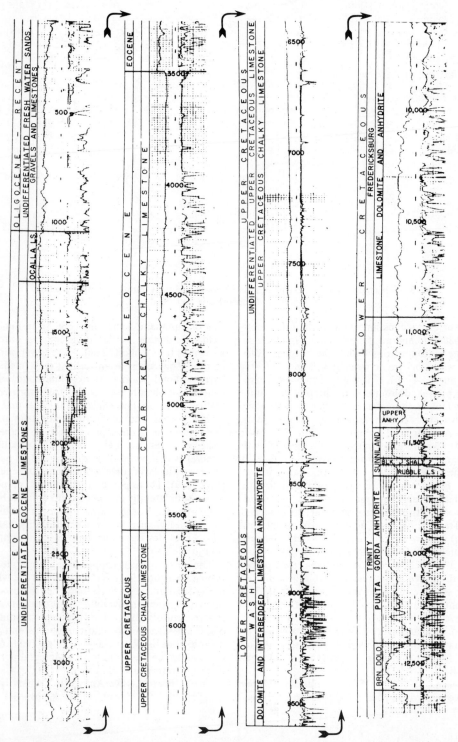

FIG. 4—Composite log showing age and general lithology of sedimentary section at Sunoco-Felda field.

"Roberts Zone." Argillaceous limestones interbedded with dolomites and anhydrites compose the approximately 400 ft (120 m) of upper Trinity section from the top of the "Upper Anhydrite" to the top of the Trinity, which lies at about 10,930 ft (3,330 m). Strata of Fredericksburg age, consisting of about 1,330 ft (405 m) of limestones interbedded with thin dolomites, shales, and a few anhydrite beds, directly overlie the Trinity. Near the top of the Fredericksburg, at approximately 9,600 ft (2,925 m), noncommercial asphaltic oil shows are found occasionally. This fact, coupled with the existence of excellent porosity and permeability in some of the Fredericksburg limestones, indicates that this zone has oil-producing potential elsewhere on the shelf. The section extending from the Fredericksburg to the top of the Lower Cretaceous is predominantly dolomite interbedded with thin limestones, anhydrites, and a few shale beds. The top of the Lower Cretaceous, at about 8,400 ft (2,560 m), is identified by the presence of a bed of waxy, dark green shale, 3–6 ft (1–2 m) thick.

The Upper Cretaceous section, consisting of approximately 3,100 ft (945 m) of uniformly soft, chalky limestone, is found at about 5,300 ft (1,615 m). This interval commonly drills at the rate of 100 ft (30 m) per hour, under optimum conditions.

Above the Cretaceous is the thick Cedar Keys Limestone of Paleocene age, which lies at about 3,500 ft (1,067 m). This soft, chalky, vuggy limestone, interbedded with hard dolomites and limestones, is the time-equivalent of the Midway Group of the Western Gulf area (Puri and Vernon, 1964).

Undifferentiated Eocene carbonate rocks, topped by the cream-colored, porous Ocala Limestone, overlie the Cedar Keys. This interval is about 2,400 ft (730 m) thick, and the top of the

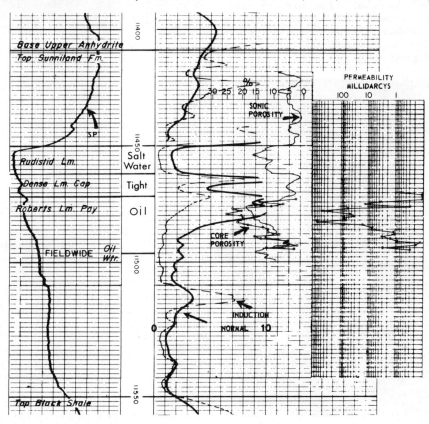

FIG. 5—Induction-electric log with sonic-log porosity and core permeability plots of "Roberts" limestone producing zone in typical Sunoco-Felda field well.

Ocala Limestone is about 1,100 ft (335 m) below the surface. These units are principally hard limestones and dolomites interbedded with soft, chalky, argillaceous limestones. In the depth interval of 2,000–3,000 ft (610–915 m), soft, chalky limestones are commonly leached out, leaving a section of caverns and ledges (Fig. 10). This 1,000-ft (300 m) interval, called the "Boulder Zone" by the oil industry, is the most hazardous interval of the entire section to drill. The drilling difficulty is caused primarily by the numerous voids and ledges encountered. There is a great variation in the size of the voids; the largest encountered in the Sunoco-Felda No. 32-2 well was 90 ft (27 m) from roof to floor. During drilling operations, the ledges form traps for the accumulation of broken pieces of formation, which may wedge between the drill pipe and a protruding ledge, causing the drill pipe to stick and twist off. Attempts to recover the parted drill pipe commonly meet with failure and the hole must be redrilled. The bottom of the "cavernous" zone is normally at about 3,000 ft (915 m), and protection casing is set at about 3,500 ft (1,065 m) to seal off this section.

FIG. 7—Typical bioclastic grainstone facies of "Roberts Zone" (×3).

Unconsolidated sands, gravels, and carbonate muds of Oligocene to Holocene age extend from the top of the Ocala Limestone to the surface. These beds contain brackish to fresh water which is protected from contamination by the surface pipe.

FIELD GEOLOGY AND RESERVOIR CHARACTERISTICS

The "Roberts" producing zone of the Sunniland Limestone in Sunoco-Felda field has a 34-ft (10 m) oil column between the oil-water contact at −11,444 ft (−3,488 m) and the highest structural point of the producing zone. Porosities and permeabilities in the reservoir range up to 28 percent and 665 md, respectively. Development drilling was essentially complete by July 1966, when 25 productive oil wells had been drilled on a spacing pattern of 160 acres. Currently there are 17 active oil wells. All but one of the abandoned oil completions have been converted to saltwater-injection wells. Approximately 4,500 productive acres (7 km²) lie within the confines of the field.

The structural configuration of the field is shown in Figures 11 and 12. The downdip limit of oil accumulation is governed by the oil-water contact, and the updip limit is formed by a facies change from the porous and permeable limestone

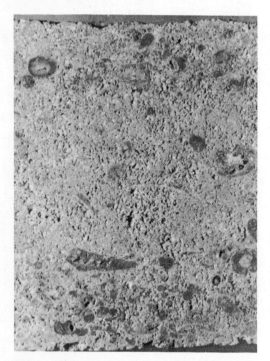

FIG. 6—Algal-plate–gastropod carbonate bank typical of "Roberts Zone" oil reservoir (×1).

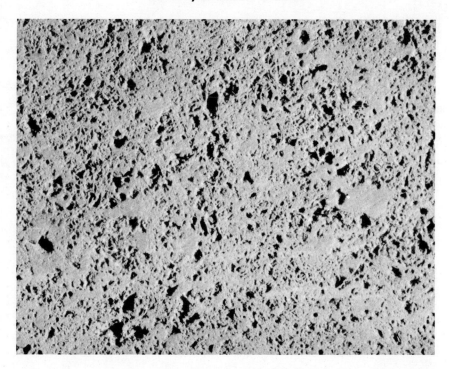

FIG. 8—Core slab of "Roberts" limestone producing section composed of gastropods, algal plates, rudistid fragments, and bioclastic debris in a recrystallized limestone matrix (×3).

reef to a chalky, miliolid-bearing, impermeable limestone. The best porosity is found basinward on the reef front, indicating that perhaps wave action and a higher level of energy played a part in forming the excellent porosity present in the reservoir rock.

Reservoir conditions are considerably poorer in the westernmost of the two Sunoco-Felda reef pods. Limited permeabilities and lower porosities in this area indicate a more complicated biostromal fabric.

RESERVOIR DATA

Reservoir data for the "Roberts" limestone oil reservoir and other field information are summarized in Table 1.

The crude oil produced at Sunoco-Felda field is a black paraffinic-naphthenic oil containing 3.084 lb of salt per barrel and 3.42 percent sulfur. Although the oil is highly contaminated with salt, this does not appear to affect either production or equipment adversely. The oil contains an unusually small amount of gas in solution, and the average well has an initial potential gas/oil ratio of approximately 100 cu ft/bbl. Initial production

rates from individual wells range up to 427 BOPD.

DRILLING AND COMPLETION

A typical Sunoco-Felda well is spudded with a 26-in. (66 cm) hole and drilled to a depth of 100 ft (30 m) using a light-weight, fluffy mud with lost-circulation material to hold back the surface sands. Twenty-inch (50.8 cm) casing is set at this point and cemented to the surface to protect freshwater zones and prevent erosion and caving around the surface location. The same type of mud is used for drilling a 17 ½-in. (44.5 cm) hole into the top of the Ocala Limestone at approximately 1,100 ft (335 m), and 13 ⅜-in. (33.9 cm) casing is set. This casing string is also cemented to the surface to protect all freshwater zones and prevent the shallow sand and gravel beds from sloughing into the hole when circulation is lost while drilling the "Boulder Zone."

A 12 ¼-in. (31.1 cm) hole is then drilled approximately 3,500 ft (1,065 m), penetrating completely the cavern-and-ledge interval of the Eocene limestones. Use of a special sealed-bearing tungsten-carbide bit for drilling this hazardous in-

Fig. 9—Core slab of "Roberts Zone," algal-plate–gastropod-bearing limestone showing excellent vuggy porosity (×3).

terval generally makes it possible to penetrate the entire zone without changing bits. Another casing string (9⅝-in. or 24.4 cm) is run and cemented to 3,500 ft. An 8¾-in. (22.2 cm) hole is drilled to the core point with constant addition of fresh water into the drilling-fluid system to compensate for the inability of the open hole to hold a full column of water. Because of lost-circulation problems in drilling the section below the 9⅝-in. casing, well-sample recovery of the section from this point to the "Roberts Zone" is poor to nonexistent.

Most of the wells in the Sunoco-Felda field were completed in open hole after a minimum penetration of the oil "pay" with the core barrel. In these wells, a string of 5½-in. (13.9 cm) casing was cemented at the top of the porous zone and completion was made from the open hole after a light acid treatment. Later completions were made in a more conventional manner. After coring and drilling through the producing interval, 5½-in. casing was cemented on bottom and perforated. Although the early wells would flow initially, production rates were inadequate and rod pumping facilities were installed immediately

upon completion. At present, the installation of downhole centrifugal pumps is under way for those wells capable of producing large volumes of fluid.

Unitization of the royalty interests was concluded on October 1, 1968 (Sun is the only operator in the field), and the field is currently being produced under a water-injection pressure-maintenance program in an effort to combat declining bottomhole pressures. Both produced and extraneous water are injected into the oil reservoir at an average rate of 10,000 bbl/day. Response to the injection program has created a rise in working fluid levels in some of the producing wells, resulting in shallower pump depths and a more efficient operation. It is too early in the life of the waterflood to determine its effectiveness, but the response to date is encouraging.

Conclusion

The Sunoco-Felda field reef trend of the Sunniland Limestone along the ancient shoals of the South Florida shelf offers great potential for the discovery of future oil reserves. Sunoco-Felda field is an example of the type, size, and quality of

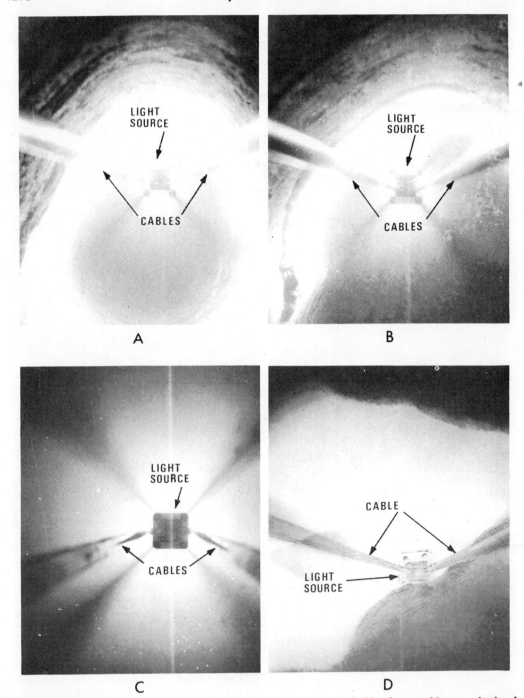

FIG. 10—Downhole photographs of "Boulder Zone" taken in clear water looking downward between depths of 2,400 and 2,500 ft (732–762 m). **A.** All sides of egg-shaped hole visible. **B.** Only one-half side of enlarged hole visible within light range. **C.** No hole sides visible within light range; camera and light source opposite cavern. **D.** Irregular hole with protruding limestone ledge.

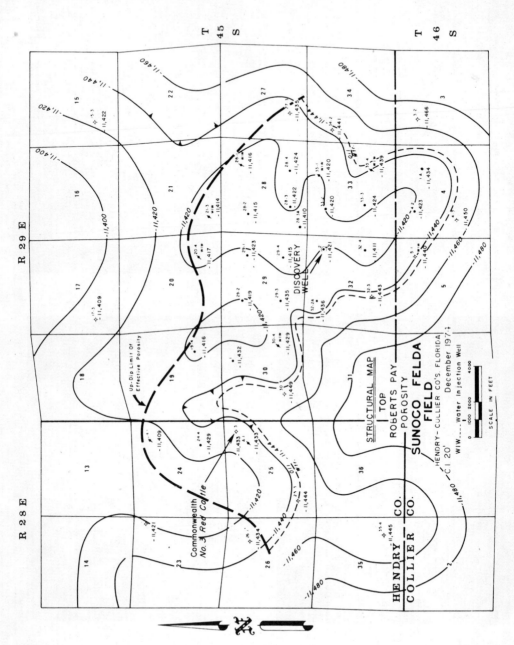

Fig. 11—Structure of top of "Roberts" limestone porous zone, Sunoco-Felda field. C.I. = 20 ft.

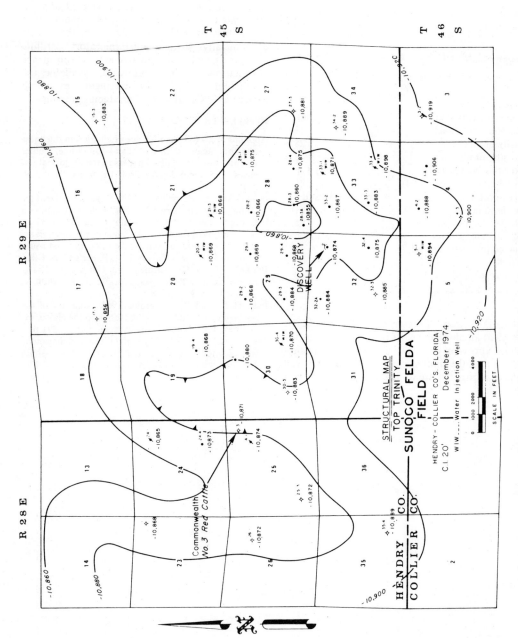

FIG. 12—Structure of top of Trinity, Sunoco-Felda field. C.I.=20 ft.

**Table 1. Reservoir and Field Data,
Sunoco-Felda Field**

Proved acreage (ac)	4,500
Well spacing (ac)	160
No. producing wells (orig.)	25
Av. depth to producing zone (ft)	11,475
Av. producing-zone thickness (ft)	13
Av. porosity (%)	18
Av. permeability (md)	60
Av. water saturation (%)	38
Gravity of oil ($^{\circ}$API)	25
Viscosity of oil (cp)	3.0
Formation volume factor	1.1/1
Gas-solution ratio (cu ft/bbl)	88
Type of drive	Limited bottom water
Orig. bottomhole pressure (psia)	5,166 @ -11,421 ft
Reservoir temperature ($^{\circ}$F)	195
Orig. oil in place (bbl)	44,000,000
Cum. oil production to 1-1-75 (bbl)	7,400,000
Av. daily oil production (bbl)	1,700

stratigraphic-trap oil reservoirs that should be prevalent all along the shelf proper. However, the subtle expression of these patch reefs in the subsurface and the limited amount of geologic well control along the trend make locating them a high-risk venture. The entire shelf area is very sparsely drilled, but the use of complete and detailed stratigraphic and environmental subsurface studies, coupled with carefully selected and sophisticated geophysical programs, should make exploration for additional reserves of this nature economically attractive. Exploratory drilling along the trend is fairly active and, as subsurface information becomes more readily available, the South Florida shelf and embayment area should emerge as a significant oil-producing province.

REFERENCES CITED

Applin, P. L., 1960, Significance of changes in thickness and lithofacies of the Sunniland Limestone, Collier County, Florida, *in* Short papers in the geological sciences: U.S. Geol. Survey Prof. Paper 400-B, p. 209-211.

—— and E. R. Applin, 1944, Regional subsurface stratigraphy and structure of Florida and southern Georgia: AAPG Bull., v. 28, no. 12, p. 1673-1753.

—— and —— 1965, The Comanche Series and associated rocks in the subsurface in central and south Florida: U.S. Geol. Survey Prof. Paper 447, p. 1-84.

Banks, J. E., 1960, Petroleum in Comanche (Cretaceous) section, Bend area, Florida: AAPG Bull., v. 44, no. 11, p. 1737-1748.

Puri, H. S., and R. O. Vernon, 1964, Summary of the geology of Florida and a guidebook to the classic exposures: Florida Geol. Survey Spec. Pub. 5, 291 p.

Sears, S. O., 1974, Facies interpretations and diagenetic modifications of the Sunniland Limestone, south Florida: Southeastern Geology, v. 15, no 4, p. 177-191.

Copyright © 1972 by The American
Association of Petroleum Geologists

The American Association of Petroleum Geologists Bulletin
V. 56, No. 8 (August 1972), P. 1419-1447, 21 Figs.

Depositional Environments and Geologic History of Golden Lane and Poza Rica Trend, Mexico, an Alternative View[1]

A. H. COOGAN,[2] D. G. BEBOUT,[3] and CARLOS MAGGIO[4]
Kent, Ohio 44240, Houston, Texas 77001, and Baytown, Texas 77520

Abstract Middle Cretaceous cores from the prolific oil fields of the Golden Lane and the Poza Rica trend in eastern Mexico were studied to determine the environment of deposition of the reservoir and associated rocks, to consider the significance of sedimentary facies for interpreting the geologic history of the Tampico embayment, and to compare the middle Cretaceous carbonate rocks and history of this area with others of the same age along the Texas Gulf Coast.

The Golden Lane fields produce from the El Abra Limestone, which was deposited in a shallow-water shelf or lagoon with scattered rudist patch reefs. The structurally lower Poza Rica trend fields contain rocks of the Tamaulipas and Tamabra Limestones. The Tamaulipas Limestone was deposited principally under open-marine, basinal conditions. The Tamabra Limestone is composed of shallow-water coral-rudist reefs, debris derived from the reefs and deposited in shoal-water nearby, and forereef talus mixed with basinal muds. Production in the Poza Rica trend is mainly from the reef debris. No coral-rudist reef was recognized in the small amount of available core examined from the Golden Lane, and present data do not support the prevalent view that the Golden Lane is a barrier reef, or reef-fringed atoll, or that the Tamabra Limestone represents deep-water deposits transported 8-16 km (5-8 mi) from the supposed Golden Lane barrier reef.

The carbonate rocks of the Golden Lane and Poza Rica trend and of the "Deep Edwards" trend in south Texas are approximately the same age and, broadly speaking, were deposited under similar environmental conditions on a shallow shelf and at the shelf edge adjacent to a basin. However, the Golden Lane and Poza Rica trend are only about 60 km (37 mi) from the Sierra Madre Oriental, a major early Tertiary orogenic belt, whereas the "Deep Edwards" trend is hundreds of miles from the same belt. Movements associated with the early Tertiary orogeny caused exposure and subaerial leaching, producing remarkable porosity in the Golden Lane. Thus, although depositional environments of the middle Cretaceous in south Texas parallel those of eastern Mexico, the subsequent geologic histories of the two regions are markedly different.

Introduction

Middle Cretaceous carbonate reservoirs of the Golden Lane and Poza Rica trend in northeastern Veracruz, Mexico, have been the source of more than 2 billion bbl of oil. The purpose of this study is to describe and delineate the distinctive facies in these two oil-prolific trends, to compare them with similar well-known facies of the much less productive Cretaceous "Deep Edwards" trend of the Texas Gulf Coast, and to add new information and interpretation of the Tampico embayment fields that will increase our understanding of the petroleum geology of one of North America's "giants."

The older parts of the Golden Lane were discovered more than a half century ago; the Poza Rica field was discovered in 1930. The interpretation of the Golden Lane as a reef, later as a barrier reef, is based originally on the summaries of scanty lithologic data by Muir (1936). It was not until the 1950s that an attempt was made to describe the rocks of the Tamabra Limestone, the reservoir for the Poza Rica and other fields, when Barnetche and Illing (1956) published their paper interpreting a part of the Tamabra Limestone as "rudist shoals." The prevailing view that the Tamabra Limestone represents slump, slide, or turbidite deposits contradicts Barnetche and Illing's (1956) conclusion. The difference is an important one. Whichever view is correct, the implications

[1] Manuscript received, September 7, 1971; accepted, November 10, 1971. Presented at the 54th annual meeting of the Association, Dallas, Texas, April 15, 1969. Published by permission of Esso Production Research Co., Houston, Texas. Kent State University, Department of Geology Contribution no. 72.

[2] Kent State University.

[3] Esso Production Research Co.

[4] Humble Oil & Refining Co.

Nearly 400 m (1,300 ft) of core from 14 wells, 10 in the Poza Rica trend and 4 on the Golden Lane platform, were obtained from the Poza Rica District Office of Pemex. A structure contour map, cross sections, mechanical logs, and other data were provided through the courtesy of Pemex geologists and management. Materials from the "Deep Edwards" trend were provided for comparison by district offices in the Exploration Department of the Humble Oil and Refining Company.

Cordial cooperation was received from the many Pemex geologists who provided material for this study including Eduardo J. Guzmán and Federico Mina-U. in Mexico City; Antonio Acuña, Francisco Acevedo, Efraín Capriles, Roberto Hernández, Homero Najera, Francisco Medina-Mora, and Armando Rentería in Poza Rica; Edmundo Cepeda, Rodolfo Suárez, and José Carillo in Tampico. Humble Oil and Refining Company geologists C. W. Holcomb, D. C. Edwards, and Hunter Yarborough aided in obtaining the cores. Thin sections were examined for datable foraminifers by the late N. K. Brown, Jr., Esso Production Research Company.

Eduardo J. Guzmán, Instituto Mexicano del Petróleo, and Paul Enos, State University of New York, Binghamton, New York, read the paper and offered valuable comments.

for the sedimentological and structural history of the Golden Lane and Poza Rica trend are immense. We began our study of cores loaned by Petróleos Mexicanos (Pemex), accepting the view of the reefal origin of the El Abra Limestone in the Golden Lane and deep-water deposition of the Tamabra Limestone; we became convinced of the opposite view, slowly, reluctantly, and against considerable opposition, but nevertheless were convinced by the preponderance of evidence in the cores and its coordination with the regional cross sections provided by Pemex. That evidence is presented here, with an emphasis on the carbonate facies of the Golden Lane and Poza Rica trend.

REGIONAL SETTING

The Golden Lane and Poza Rica trend lie near the center of the Tampico embayment, a major structural feature of eastern coastal Mexico (Fig. 1). The embayment is bounded on the north by the Tamaulipas arch, on the south by the Jalapa uplift, and on the east by the Golden Lane platform, a constructional feature. West of the Tampico embayment is the Sierra Madre Oriental range, which developed as a result of the Laramide orogeny. Between the Sierra Madre Oriental and the Golden Lane platform (the term we prefer to barrier reef trend or atoll) lies the narrow Chicontepec basin, primarily a Tertiary feature (López-Ramos, 1959).

A partial double arc consisting of the Golden Lane on the inside (now gulfward) and the Poza Rica trend on the outside (now landward) curves southwest from Cabo Rojo, about 80 km (50 mi) south of Tampico on the Gulf of Mexico, to a point 40 km (25 mi) west of Tuxpan, and then southeastward back to the coast, east of the Poza Rica field (Fig. 2). The Golden Lane arc was interpreted as a single, long, narrow barrier reef until seismic surveys in the early 1960s indicated closure of the structural trend in the Gulf of Mexico. The Golden Lane fields now are known to lie along the rim of a northwest-southeast-trending oval structural platform, a so-called "atoll" (Guzmán, 1967; Viniegra and Castillo-T., 1968), having approximate axes of 145 km (90 mi) by 65 km (41 mi). This platform and the structurally lower Poza Rica trend, which partly parallels the western side of the platform, dip southeastward into the Gulf of Mexico.

STRATIGRAPHY

The Lower and middle Cretaceous carbonate rocks of the Tampico embayment are underlain by mainly shallow-water, marine Jurassic limestone, and older redbeds, schists, and igneous

rocks (Fig. 3), and are overlain by a thin section of open-marine Upper Cretaceous shale and limestone and a thick Tertiary section of sandstone and shale, representing a basin-filling molasse. Eocene, Oligocene, and younger Tertiary strata crop out in the Tampico embayment. Around the margins of the embayment, Upper Cretaceous rocks crop out on the Tamaulipas arch on the north; Upper and middle Cretaceous rocks crop out on the northwest in the Sierra del Abra and on the west in the foothills of the Sierra Madre Oriental; and Precambrian, Paleozoic, and Mesozoic rocks crop out in the Sierra Madre Oriental.

Three formations of the middle Cretaceous of the Golden Lane and Poza Rica trend are considered primarily in this report: the El Abra, Tamabra, and upper Tamaulipas Limestones (Fig. 4). Each of these comprises a variety of rock types and each has been interpreted to have been deposited in a different environment.

The El Abra Limestone, named for outcropping rocks in the Sierra del Abra, in the subsurface consists of interbedded rudist, miliolid, oolitic, and micritic limestone; it has been interpreted (Muir, 1936; Rockwell and García-Rojas, 1953; Boyd, 1963; Guzmán, 1967; Viniegra and Castillo-T., 1970) along the Golden Lane as a barrier reef or reef-fringed atoll with the center of the oval Golden Lane platform interpreted as a lagoon with evaporitic deposits. The El Abra Limestone is reported from early records to be as thick as 2,537 m (8,118 ft) in the Jardín No. 35 well, and the thickness is estimated by Viniegra and Castillo-T. (1970) to be 1,467 m (5,108 ft). However, the information is from a questionable drilling record, possibly not a vertical hole (Viniegra and Castillo-T., 1968; Guzmán, 1967) and in our view should not be considered pertinent. Guzmán (1967), nevertheless, supported the view that the El Abra could be as thick as 2,000 m (6,500 ft), but the deepest confirmed penetration in the Golden Lane is 1,323 m (4,233 ft) in the Arroyo Grande No. 1. In the central part of the Golden Lane platform the El Abra Limestone averages 1,400 m (4,480 ft) according to Guzmán, (1967). Viniegra and Castillo-T. (1970) reported the El Abra Limestone to be 762 m (2,509 ft) thick in the Mesita No. 1 and 1,174 m (3,840 ft) thick in the Cañas 101; both New Golden Lane wells penetrated the entire formation. The top of the El Abra Limestone ranges in depth from about 500 m (1,640 ft) below sea level at the middle of the western side of the Golden Lane arc to more than 2,500 m (8,200 ft) at the southeastern end (Fig. 5). Considerable relief on the upper surface of the formation (Figs. 5, 6) is the result of post-middle Cretaceous surface weathering. The relief also has been attributed to reef

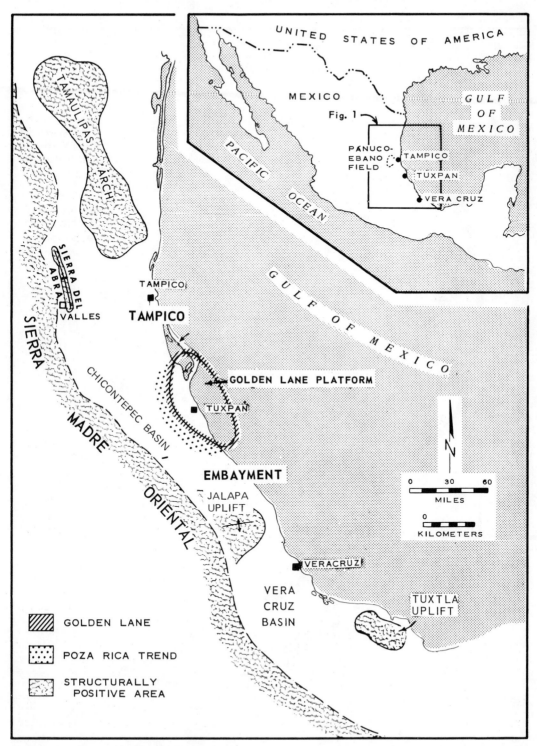

Fig. 1—Regional setting of Tampico embayment and location of Golden Lane and Poza Rica trend in eastern coastal Mexico (modified from Murray and Krutak, 1963).

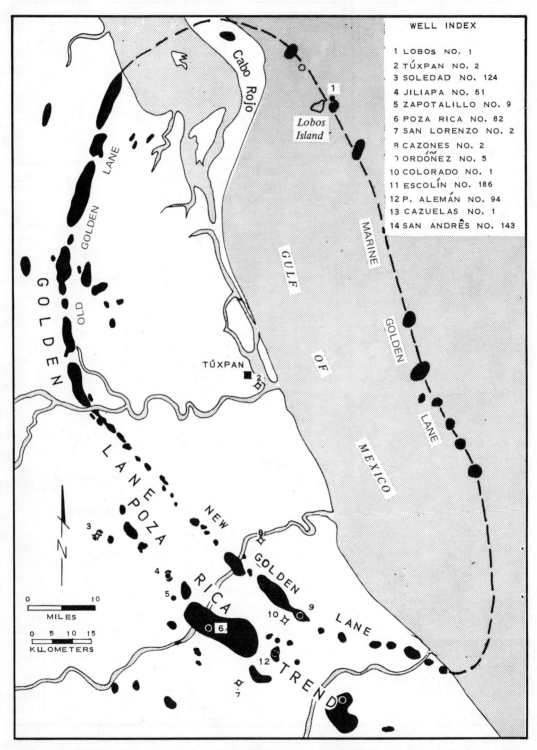

WELL INDEX

1 LOBOS NO. 1
2 TÚXPAN NO. 2
3 SOLEDAD NO. 124
4 JILIAPA NO. 61
5 ZAPOTALILLO NO. 9
6 POZA RICA NO. 82
7 SAN LORENZO NO. 2
8 CAZONES NO. 2
9 ORDÓNEZ NO. 5
10 COLORADO NO. 1
11 ESCOLÍN NO. 186
12 P. ALEMÁN NO. 94
13 CAZUELAS NO. 1
14 SAN ANDRÊS NO. 143

Fig. 2—Oil fields of Golden Lane and Poza Rica trend. Cores studied are from wells named on map.

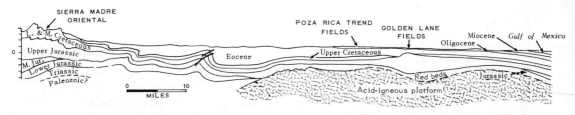

(AFTER LÓPEZ RAMOS, 1959.)

FIG. 3—Regional cross section from Gulf of Mexico across Golden Lane and Poza Rica trend to Sierra Madre Oriental. Pre-Jurassic acid-igneous basement underlies trends and presumably is base for constructional Golden Lane platform. (Modified from López-Ramos, 1959.)

growth or faulting normal to the trend or to both. The El Abra Limestone in the Golden Lane is overlain by rocks ranging in age from Late Cretaceous to Oligocene; in the central part of the Golden Lane platform a thin but apparently continuous sequence of Upper Cretaceous and Tertiary deposits blankets the El Abra Limestone.

The Tamabra Limestone of the Poza Rica trend, named from a combination of the words Tamaulipas and El Abra, is composed of bioclastic, detrital, rudist, and micritic-planktonic limestones, which have been interpreted (Bonet, 1952, 1963; Boyd, 1963; Becerra, 1970) as forereef and turbidite deposits interbedded with basinal sediments. Age determinations based on microfossils from the micritic-planktonic limestone range from early Albian to Cenomanian, according to the late Noel K. Brown, Jr.

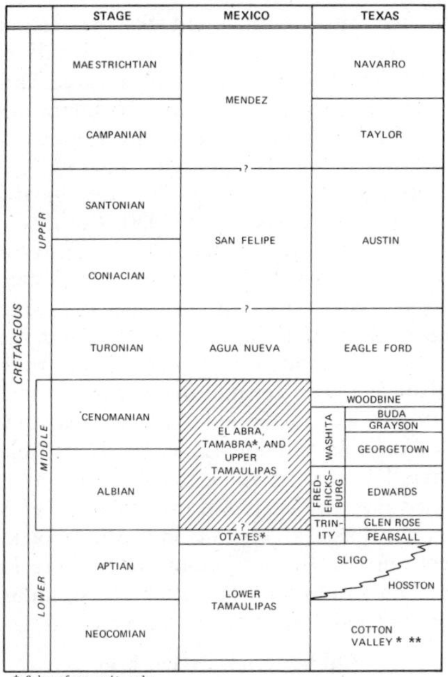

The Tamabra Limestone has been penetrated only in the Poza Rica trend west of the Golden Lane, where it is about 1,000 m (3,280 ft) structurally lower than the top of the equivalent El Abra Limestone (Fig. 6). The Tamabra Limestone decreases in thickness from more than 200 m (656 ft) near the Golden Lane to a featheredge on the west and southwest (Fig. 7), where it pinches out between the underlying middle Cretaceous upper Tamaulipas Limestone and the overlying Late Cretaceous, mainly Turonian, Agua Nueva Formation.

The upper Tamaulipas is predominantly a micritic-planktonic limestone and is present westward from the Poza Rica trend to the Sierra Madre Oriental. It is considered a basinal sediment. Brown dated this limestone in the Poza Rica trend wells as Albian to Cenomanian on the basis of planktonic fossils. The Tamaulipas Limestone is recognized by its lithology, stratigraphic position, age, and lack of a bioclastic carbonate facies.

Problems critical to environmental interpretations are the age of the El Abra and Tamabra Limestones at their respective tops and the relation of the Agua Nueva Formation to the Tamabra Limestone. In this study, the youngest dates for the El Abra Limestone are Albian-Cenomanian in the E. Ordóñez No. 5, determined from large foraminifers and caprinid rudists, and at the top of the Lobos No. 1, determined from miliolid foraminifers. In the Tuxpan No. 2, the El Abra Limestone is overlain by planktonic-micritic facies of the Agua Nueva Formation which is Cenomanian, not Turonian as usually assumed. The upper part of the Tamabra Limestone has been

	STAGE	MEXICO	TEXAS	
CRETACEOUS / **UPPER**	MAESTRICHTIAN	MENDEZ	NAVARRO	
	CAMPANIAN		TAYLOR	
		?		
	SANTONIAN	SAN FELIPE	AUSTIN	
	CONIACIAN			
		?		
	TURONIAN	AGUA NUEVA	EAGLE FORD	
MIDDLE	CENOMANIAN	EL ABRA, TAMABRA*, AND UPPER TAMAULIPAS	WOODBINE	WASHITA
			BUDA	
			GRAYSON	
			GEORGETOWN	
	ALBIAN		EDWARDS	FREDERICKSBURG
		?	GLEN ROSE	TRINITY
		OTATES*	PEARSALL	
LOWER	APTIAN		SLIGO	
		LOWER TAMAULIPAS	HOSSTON	
	NEOCOMIAN		COTTON VALLEY * **	

* Subsurface unit only.
** In part Jurassic.

FIG. 4—Generalized correlation chart of Cretaceous rocks of Tampico embayment, Mexico, and of Texas Gulf Coast.

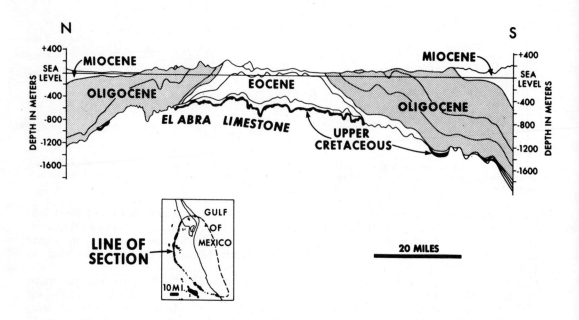

Fig. 5—Longitudinal section along Golden Lane from Cabo Rojo to New Golden Lane (Fig. 2) showing relief of more than 250 m (820 ft) on surface of El Abra Limestone and thickness of overlying Cretaceous and Tertiary formations. (Modified from Sotomayor-C., 1954.)

dated by Noel K. Brown, Jr., as Cenomanian or older, mainly from planktonic microfossils. No Turonian rudists were found in the Tamabra Limestone. On the other hand, Becerra (1970) dated the upper part of the Tamabra as Turonian, especially in parts of the Poza Rica field. Interestingly, the same micritic interval in the Zapotalillo No. 9, well below the top of the Tamabra, was dated by Brown as Cenomanian and by Becerra as Turonian. Hence, the question of the age of the top of the Tamabra remains unresolved to the detail desired. The Agua Nueva Formation in the Poza Rica trend is commonly considered Turonian. If it is interbedded with the Tamabra Limestone as shown in the cross section (Fig. 7), the Tamabra Limestone presumably would be Turonian at the top, too. The upper part of the Tamabra probably is of different ages in different places. The pertinent point is that, if it is Cenomanian, the top part is the same age as the El Abra Limestone. If it is Turonian, it is younger than the El Abra and all published correlation sections between the two formations are misdrawn. As this study showed no faunal evidence of a Turonian age for the Tamabra, the traditional view and correlations are used.

HYDROCARBON PRODUCTION AND RESERVOIR CHARACTERISTICS

The Tampico embayment has an estimated ultimate recovery of approximately 5 billion bbl of oil. Earliest production from fractured Tamaulipas Limestone was in the Pánuco-Ebano region (Fig. 1) in 1904. The part of the Golden Lane now known as the Old Golden Lane (Fig. 2) was discovered in 1908; drilling sites were located in the vicinity of oil seeps. In 1930, the discovery well of the Poza Rica field was drilled over a large gravity high. On the basis of seismic and gravity evidence, the Golden Lane was extended southeastward in 1952, and the southern extension was called the New Golden Lane. In 1958, on subsurface and seismic information, the Poza Rica trend was extended spottedly northward, west of the Old Golden Lane, partly completing two parallel arcs. In 1963, the Golden Lane was extended offshore eastward into the Gulf of Mexico on the basis of seismic indications of structural closure on the surface of the El Abra Limestone. The production statistics are summarized from published and unpublished Pemex data.

Golden Lane Production

About 50 fields of the Golden Lane produce primarily from highs in the El Abra Limestone (Fig. 5) along a narrow band 0.3-2.7 km wide (0.2-1.6 mi; Viniegra and Castillo-T., 1970). The

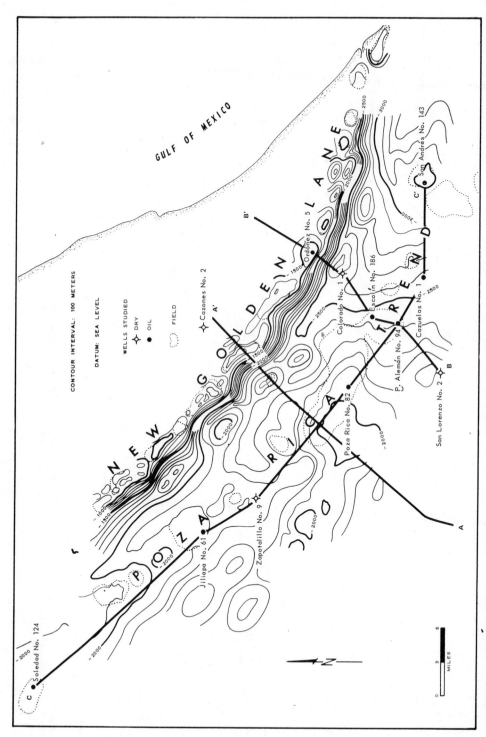

FIG. 6.—Structure contour map, datum on top of El Abra, Tamabra, and Tamaulipas Limestones and locations of wells from which core was studied. Note approximate 1,000 m (3,280 ft) of relief between top of El Abra and top of Tamabra Limestone. Lines are for cross sections illustrated in Figures 7, 9, 10, 11. (Modified from map prepared by H. Najera, G. Obregón, and F. Acevedo, Poza Rica district, Pemex.)

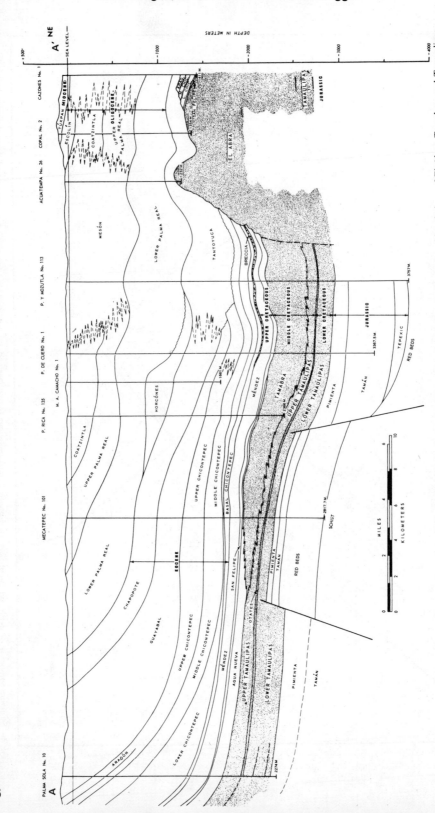

Fig. 7.—Cross section from Golden Lane platform west across Poza Rica trend showing relative thicknesses and general stratigraphic relations of El Abra, Tamabra, and Tamaulipas Limestones. Evidence for interfingering of Tamabra Limestone and Agua Nueva Formation, as shown, is scant. Location of cross section on Figure 6. (Modified from Pemex cross section prepared by P. Carillo, R. Rocha, and A. Acuña.)

Old Golden Lane reached a peak production of 455,000 bbl/day of oil in 1921 and still produced 30,000 bbl/day in 1967 (Guzmán, 1967). An early well, the Cerro Azul No. 4, once produced an estimated 260,000 bbl/day. Cumulative production to the end of 1967 is near 1.25 billion bbl.

The New Golden Lane consists of 17 fields ranging in width from 1.5 to 7 km (0.9–4.5 mi, average 1.3 sq km or 0.8 sq mi) and has estimated reserves of 350 million bbl (Guzmán, 1967). Oil and gas in the New Golden Lane fields are produced from highs, especially elongate highs, in the El Abra Limestone which is directly overlain by Upper Cretaceous, Eocene, or Oligocene (Fig. 8). Gravity of the crude oil ranges from heavy (API 14°, Galeana and Morelos fields) to semi-heavy (API 20°, Santa Agueda, and E. Ordóñez fields) to light (API 30°, Guerrero and Hidalgo fields). The first five wells drilled in the E. Ordóñez field, developed on 64-ha. (160 acre) spacing, had an average initial production of 8,100 bbl/day from an oil column up to 60 m (200 ft) thick. Oil is produced from different levels in separate New Golden Lane fields. The source of the oil is thought by some to be the Tertiary fine-grained clastic rock, other geologists consider the source of the oil to be Jurassic black shale and limestone, and others believe that the oil might be indigenous to the middle Cretaceous limestones.

Marine extensions of the Golden Lane first were drilled for new field discoveries directly off Cabo Rojo at Lobos Island (Fig. 2) and just southeast at Arrecife Medio. Since 1964, nine other commercial Marine Golden Lane fields have been discovered. The Tiburón, Esturión, Bagre, and Atún fields together produce more than 19,000 bbl/day of oil and 21,200 Mcf/day of gas (Díaz-González, 1969; World Oil, 1969). The Foca, Pez Vela, Pargo, and Morsa fields produce light, clean oil at an average rate of 3,000 bbl/day (World Oil, 1969; Franco, 1969). The Escualo structure, drilled in 1969, produced an initial 1,990 bbl/day (Acosta-Estévez, 1970). The cumulative production of all Golden Lane fields to the end of 1967 was more than 1.42 billion bbl of oil (Viniegra and Castillo-T., 1968).

Original intergranular porosity in the New Golden Lane wells is in bioclastic-rudist limestones and in the miliolid calcarenites of the El Abra Limestone. Solution and recrystallization also developed secondary porosity in limestones that originally had low porosity. Tremendous porosity and permeability in the upper part of the El Abra Limestone are indicated by loss of circulation in wells immediately after penetrating the

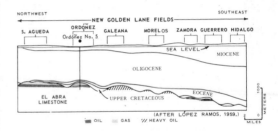

Fig. 8—Strike section along part of New Golden Lane showing oil and gas fields. Petroleum is trapped in highs in El Abra Limestone below Cretaceous and Tertiary clastic rocks. Approximate location of section is shown in Figure 6.

formation, by poor core recovery, and by records of pressure drops in wells more than 3 km (2 mi) away from new wells put on production. The pressures normally allow drilling of only the upper few feet of the El Abra Limestone. The high permeability results in a primary oil recovery as high as 60 percent in some fields of the Golden Lane. The high and variable porosity and permeability are attributed to subaerial erosion (karst development) of the El Abra Limestone during the early Tertiary, possibly as a result of eustatically lower sea level.

Several fields are within the oval Golden Lane platform at Solís and Frijolillo, where production is from structural highs in the lagoon, supposedly reef highs (Acosta-Estévez, 1969).

Poza Rica Trend Production

Poza Rica production is primarily from the Tamabra Limestone in approximately 19 fields ranging in size from 2.6 to 108 sq km (1–42 sq mi), an average of 15 sq km (6 sq mi). The Poza Rica field, the largest in the 16-km (10 mi) wide trend, produced from a third to a half of Mexico's oil in 1964 and had produced more than 650 million bbl of oil by then. In 1957, more than 200 wells in this field, which is developed on 64-ha. (160 acre) spacing, produced 100,000 bbl/day of light-gravity crude (API 34°) with a gas-oil ratio of 326. The Poza Rica trend produced 1 billion bbl of oil by June 1966 and had reserves of 2 billion bbl at the then-current production rate of 105,000 bbl/day (Guzmán, 1967).

Production in the Poza Rica field is essentially from a stratigraphic trap caused by the facies change westward from porous bioclastic Tamabra Limestone into dense planktonic-micritic upper Tamaulipas Limestone and overlying nonporous rock of the Upper Cretaceous Agua Neuva

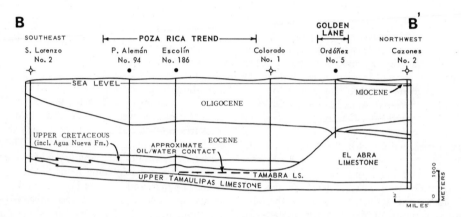

Fig. 9—Dip section from Golden Lane through Poza Rica trend. Oil in Poza Rica trend fields is trapped by westward pinchout of porous Tamabra Limestone.

Formation. Tamabra production is limited eastward by the oil-water contact (Fig. 9). Seismic data suggest to some that similar Poza Rica structures lie on the southeast side of the Marine Golden Lane. Pemex has had tentative plans to drill such structures.

Porous intervals in the Tamabra Limestone coincide with the presence of coarse and fine bioclastic and micritic-caprinid limestones. Micritic-planktonic and detrital limestones generally are not porous. Extensive recrystallization, dolomitization, and fracturing have enlarged the original intergranular porosity in the bioclastic limestones and have developed porosity in the originally nonporous matrix of the micritic-caprinid limestones. Unaltered nonporous rocks, including shales or claystones, may separate different reservoirs vertically. Effective porosity, excluding compact layers, averages 14 percent and ranges from 8 to 20 percent. Permeability is highest, exceeding 1 darcy, where alteration is greatest, and is lowest, less than 5 md, where there is no alteration. Facies change and structural position control vertical and horizontal variations of porosity.

CARBONATE FACIES

Facies Summary

Distinctive associations of lithologies, textures, structures, and fossils provided the basis for recognizing 16 facies in the cores examined from wells in the New Golden Lane, the lagoonal area on the Golden Lane platform, and the Poza Rica trend. Twelve of the facies were recorded from wells in the New and Marine Golden Lane and on the platform (Fig. 10); eight facies were recorded from wells in the Poza Rica trend (Fig. 11).

Descriptions of the more common facies, including data on pertinent petrographic details, faunas, textures, and structures, and subsequent alteration and porosity, are presented in the following section. Each facies is interpreted environmentally by the use of a combination of significant petrographic and paleontologic criteria and relations to associated facies.

Facies of the Golden Lane platform include the argillaceous-micritic, calcareous-shale, caprinid-micritic, coarse-bioclastic, coated-grain, fine-bioclastic, fragmental, laminated-algal, miliolid-calcarenite, miliolid-micritic, and molluscan-micritic. Some of these facies are illustrated in Figures 12–15.

Facies of the Poza Rica trend include the coarse-bioclastic, fine-bioclastic, caprinid-micritic, planktonic-micritic, algal-micritic, detrital,

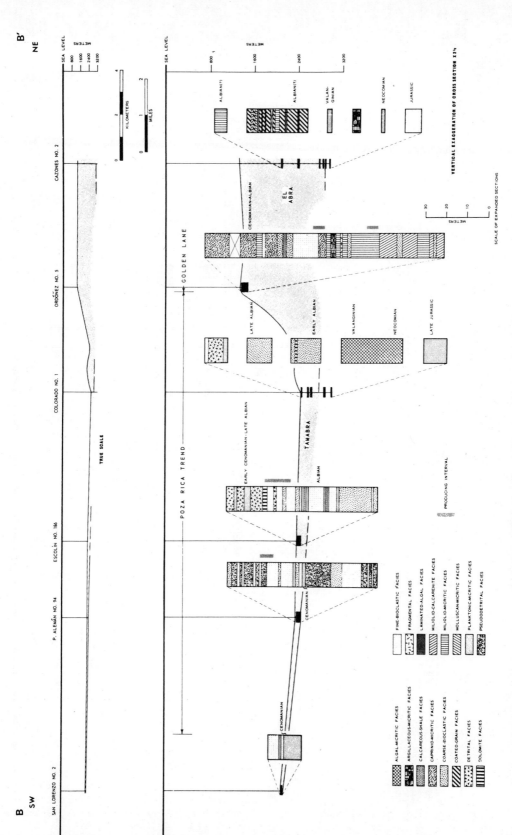

Fig. 10—Cross section from Golden Lane platform to Poza Rica trend showing locations of cores studied and their interpreted facies.

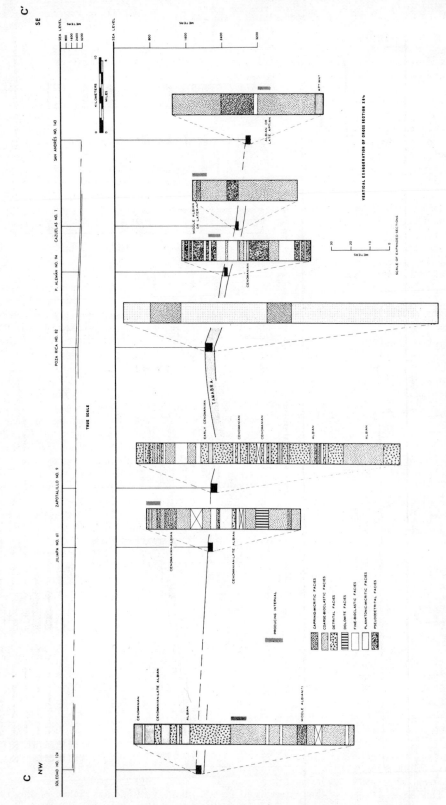

FIG. 11—Longitudinal section along Poza Rica trend showing locations of cores studied and their interpreted facies.

pseudodetrital, and dolomite. The first four facies also are present in the New Golden Lane; the last two facies are based solely on petrographic criteria and represent alteration of one or more of the other facies to such an extent that the original textures cannot be recognized. The coarse-bioclastic, fine-bioclastic, and planktonic-micritic facies are the commonest. Facies of the Poza Rica trend are illustrated in Figures 13-17.

Description of Major Facies

Miliolid-calcarenite facies (Fig. 12A)—The miliolid-calcarenite facies is characterized by light-gray skeletal limestone composed primarily of abundant tests of miliolid foraminifers; it has sparry-calcite cement and sparse micrite. Toucasid rudists and skeletal debris of other pelecypods are present in this facies. Where sparry calcite is leached or has not been deposited, the miliolid-calcarenite is very porous. The producing interval in the Lobos No. 1 is mainly in this facies. The E. Ordóñez No. 5 produced heavy oil from this facies at a depth of approximately 1,550 m (5,085 ft). In addition to being present in those two wells, the facies was penetrated in the Cazones No. 2 and in the El Abra Limestone in the Sierra del Abra. It is interpreted as a shallow-shelf or lagoonal deposit and is associated with miliolid-micritic and coarse-bioclastic facies.

Miliolid-micritic facies (Fig. 12 B, C)—Light-gray to tan micritic limestone containing sparse to common miliolid foraminifers distinguishes this facies. Irregular laminations and burrows are common structures. In addition to miliolids, other small foraminifers, ostracods, and gastropods are present.

Locally, this facies is fractured and leached. Primary intergranular porosity is lacking, but leached porosity is present in the producing interval in the Lobos No. 1, Marine Golden Lane.

The miliolid-micritic facies is interbedded with miliolid-calcarenite, bioclastic, fragmental, and molluscan-micritic limestone in the Golden Lane and area on the east; it crops out in a quarry at the type locality of the El Abra Limestone near Valles, San Luis Potosí. This facies is interpreted to represent shallow-shelf or lagoonal deposits.

Molluscan-micritic facies (Fig. 12D)—The molluscan-micritic facies is a light-gray to tan skeletal-micritic limestone containing common to abundant whole and fragmented shells of several types of mollusks; locally it is fractured and stylolitic. Miliolid foraminifers, other microfossils, and branching algae also are present in this facies. Primary intergranular porosity is lacking, but fracture porosity may be significant.

This facies makes up thin intervals in cored sections of the Tuxpan No. 2 and Cazones No. 2. It represents a shallow-shelf or lagoonal deposit.

Calcareous-shale facies (Fig. 12E)—Gray to green, quartzose, silty, calcareous-shale with irregular laminae distinguishes this facies; no fossils were found. The facies occupies a thin interval in the E. Ordóñez No. 5, where it is interbedded with the coarse-bioclastic facies, and in the Cazones No. 2, where it is associated with the molluscan-micritic facies and with oolitic limestone of the coated-grain facies. The association of this facies with the molluscan-micritic and coated-grain facies provides the basis for interpreting it as a shallow-shelf or lagoonal deposit.

Laminated-algal facies (Fig. 12F)—Light-gray micritic and micritic-skeletal limestone containing zones of dense, laminated, algal-micritic limestone are included in this facies. Irregular laminae, which characterize this facies, represent poorly perserved impressions of matlike algal growths which cemented and bound the fine-grained calcium carbonate mud. Branching dasycladacean algae and miliolid foraminifers are sparse. Vuggy porosity is well developed in the micrite between the denser algal laminations. This facies is found in the Cazones No. 2 and in the Lobos No. 1. In the Lobos No. 1 it is part of the producing interval. The laminated-algal facies is interbedded with the miliolid-micritic, miliolid-calcarenite, and coated-grain facies. It is interpreted as a shallow-shelf or lagoonal deposit which possibly formed in the intertidal zone.

Coated-grain facies (Fig. 13A, B)—The rock consists of light-gray algal-encrusted skeletal limestone, oolitic limestone, and composite-grain limestone in which coated grains are the predominant constituent. The grains, which are well to moderately well sorted, commonly are cemented by sparry calcite, but intergranular porosity may be present where calcite has been leached. The facies contains several strongly asphaltic intervals. Fossils include abundant dasycladacean algae, large gastropods, miliolids, and pelecypods.

This facies is recognized only in the Cazones No. 2, Golden Lane platform, where it is interbedded with the laminated algal, miliolid-micritic, and miliolid-calcarenite facies. It is interpreted as a shallow-shelf oolite bank with associated shoal-water deposits that formed on a current- or wave-agitated bottom.

Caprinid-micritic facies (Fig. 14A-E)—This facies includes light-gray rudist micritic limestone and rudist micritic-skeletal limestone. Caprinids are the most abundant rudists. Radiolitid rudists, colonial corals, stromatoporids, gastropods, echinoids, and foraminifers are common to abundant. The micritic matrix of this facies is not

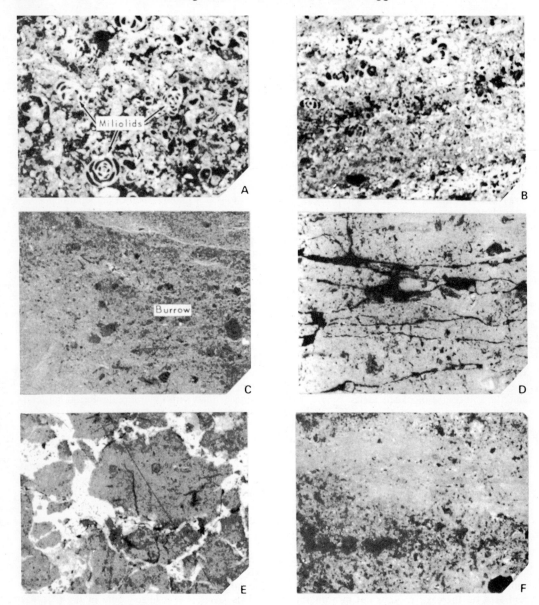

Fɪɢ. 12—Facies of Golden Lane.

A. Miliolid-calcarenite facies from producing zone of Lobos No. 1 at depth of 2,099 m (6,885 ft). Miliolid limestone, specimens of *Nummoloculina* sp. cemented by minor sparry calcite. Intergranular pore space present. Negative print of thin section, ×15.

B. Miliolid-micritic facies. Specimens of *Nummoloculina* sp. in micritic matrix. From depth of 1,530 m (5,018 ft) in E. Ordoñez No. 5. Negative print of peel, ×10.

C. Miliolid-micritic facies. Burrowed, laminated micritic-skeletal, and micritic-miliolid limestone from depth of 2,126 m (6,975 ft) in Cazones No. 2. Negative print of peel, ×3.

D. Molluscan-micritic facies. Micritic limestone containing fragments of mollusk shells and miliolid foraminifers. Abundant horizontal and vertical fractures may be desiccation cracks. From depth of 1,536 m (5,038 ft) in Tuxpan No. 2. Negative print of peel, ×3.

E. Calcareous-shale facies. Argillaceous miliolid fragmental limestone. Fragments of miliolid micritic limestone in matrix of green shale. Origin of texture unknown; it may represent storm or tidal-flat deposition. From depth of 1,523 m (5,015 ft) in E. Ordónez No. 5. Negative print of peel, ×3.

F. Laminated-algal facies. Stromatolitic-micritic limestone having vuggy porosity. From depth of 2,438 m (7,995 ft) in Cazones No. 2. Negative print of peel, ×3.

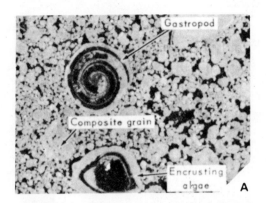
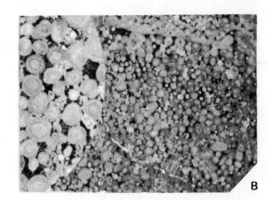

Fig. 13—Facies of Golden Lane.

A. Coated-grain facies. Encrusted algal-skeletal limestone. Skeletal grains, ooliths, and indeterminate particles coated with encrusted algal laminae. From depth of 2,442 m (8,010 ft) in Cazones No. 2. Negative print of peel, ×3.

B. Coated-grain facies. Oolitic limestone with horizontal fractures, stylolites, and sparry calcite cement. From depth of 2,430 m (7,970 ft) in Cazones No. 2. Negative print of peel, ×5; enlargement, ×15.

porous, but minor fractures may provide some porosity. This facies was penetrated in the E. Ordóñez No. 5 in the New Golden Lane, on the outcrop as the Taninul Member of the El Abra Limestone in the Sierra del Abra (Bonet, 1952, 1963), and in four wells—Jiliapa No. 61, Soledad No. 124, Cazuelas No. 1, and San Andrés No. 143—in the Poza Rica trend. In general, this facies is interpreted as a shallow-shelf rudist reef deposit.

In the caprinid-micritic facies of the Poza Rica trend, extensive recrystallization of the micritic matrix has affected as much as 30-50 percent of the rock, and dolomitization commonly as much as 70-80 percent, resulting in intervals many meters thick of highly porous carbonate rocks. Production in both the Jiliapa No. 61 and the Cazuelas No. 1 is in part from the caprinid-micritic facies.

Rudists are the framework-building organisms; colonial corals, stromatolitic algae, and hydrozoans (stromatoporids) add to the binding capacity. The Tamabra rudists, which are as much as 3-4 in. in diameter, appear to be in growth position, and lack signs of abrasion. Geopetal structures in the body cavity of the caprinid rudists show that the enclosing rock is in the same position as it was when lithified and probably is not a large transported block. This reef fauna in a micritic matrix is interpreted to represent a coral-rudist-reef deposit with the capacity to build a wave resistant reef-wall (Fig. 14A, C). Rudist bodies consisting primarily of caprinid rudists (Fig. 14B) are judged on the basis of relations in Texas outcrop and subsurface occurrences to be quiet-water reef, forereef or backreef, and lagoonal patch reef or bank deposits. Radiolitid (Fig. 14D) and caprotinid rudists may build lagoonal rudist reef and bank deposits in the intertidal zone.

The interpretation of rudist bodies as true reef depends on an understanding and documentation of the type of rudist fauna, associated coelenterate, hydrozoan, and algal fauna and flora, geometry and associated facies of the rudist body. The Taninul facies, which consists of rudist bodies in the El Abra Limestone of the Sierra del Abra, were interpreted by Bonet (1952) as banks. It is the presence of corals, matlike hydrozoans (stromatoporids), and binding algae, as well as the facies associations, which serve as the paleontologic and paleoecologic basis for our interpretation of the Tamabra caprinid-micritic facies in the San Andrés No. 143 and Soledad No. 124 as true reef front. The bulk of the caprinid-micritic facies is either part of the reef complex or represents backreef and lagoonal rudist banks. As

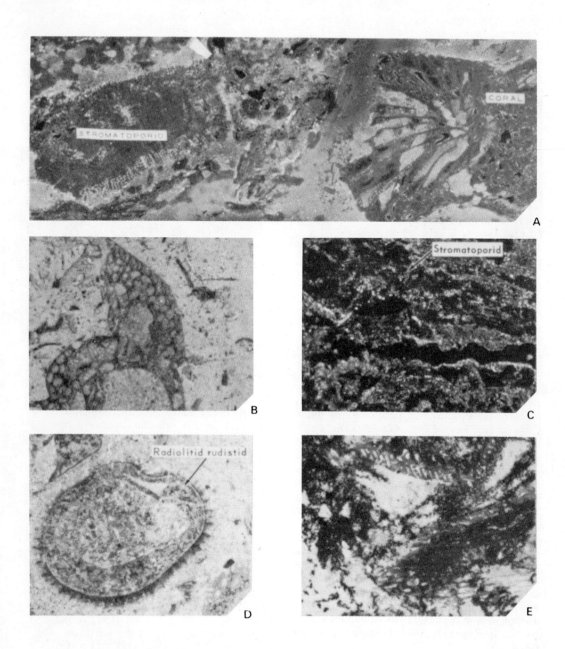

Fig. 14—Caprinid-micritic facies of Golden Lane and Poza Rica trend.

　　A. Coral stromatoporid (hydrozoan)-rudist-micritic limestone. From depth of 2,990 m (9,808 ft) in San Andrés No. 143. Negative print of peel, ×3.

　　B. Caprinid-skeletal-micritic limestone showing oblique section of whole caprinid rudist (*Caprinuloidea* sp.). From depth of 1,489-1,495 m (4,843-4,882 ft) in E. Ordóñez No. 5, New Golden Lane. Negative print of peel, ×3.

　　C. Stromatoporid (hydrozoan)-algal-micritic limestone. From depth of 2,987 m (9,798 ft) in San Andrés No. 143. Negative print of peel, ×3.

　　D. Rudist-skeletal-micritic limestone, recrystallized and asphaltic. From depth of 2,845 m (9,329 ft) in Cazuelas No. 1. Negative print of peel, ×3.

　　E. Rudist-skeletal-micritic limestone with asphaltic material in fractures and pore space. From depth of 1,950-1,955 m (6,396-6,412 ft) in Soledad No. 124. Ground surface, ×1.

might be expected, the normally narrow reef front would leave a thin deposit.

Barnetche and Illing (1956) interpreted similar facies in the Poza Rica field as rudist shoals of local origin, representing the remains of fauna that thrived in the Tamabra sea, rather than debris derived from rudist deposits of the Golden Lane. Our study supports their conclusions, at least in their most significant aspects; that is, that these deposits formed more or less in place as shallow-water, probably shallow-subtidal and intertidal, organic growth structures.

Coarse-bioclastic facies (Fig. 15A-C)—Light-gray to tan, medium-grained to gravel- and cobble-sized skeletal limestone with sparry calcite cement characterized this facies. Rudist fragments make up the most of the rock; also present are foraminifers, corals, stromatoporids, and gastropods. Leached intergranular porosity is high, especially in intervals where this facies is dolomitized and recrystallized. The coarse-bioclastic facies is oil-productive in four of the wells studied from the Poza Rica trend; it was tested in several other wells along the Poza Rica trend and in the E. Ordóñez No. 5 of the New Golden Lane.

In the Golden Lane, the coarse-bioclastic facies locally is porous. This facies in the E. Ordóñez No. 5 is interbedded with miliolid-micritic and miliolid-calcarenite facies. Similar coarse-grained bioclastic limestone crops out in a quarry of the El Abra Limestone near Valles. The associated fauna and facies suggest that the coarse-bioclastic facies in the Golden Lane was deposited on a shallow shelf as a shoal-water sediment.

In the Poza Rica trend, the coarse-bioclastic facies is typically coarser grained than that of the Golden Lane, and is fractured, stylolitic, steeply dipping, and highly altered. Many intervals are completely recrystallized and dolomitized. The increase in matrix grain size and leaching resulting from this alteration may be responsible for developing the excellent reservoir rock in the San Andrés No. 143, Soledad No. 124, Jiliapa No. 61, Cazuelas No. 1, and wells in the Poza Rica field studied by Barnetche and Illing (1956). A pseudobrecciated texture commonly accompanies this recrystallization. The coarse-bioclastic facies is interbedded with caprinid-micritic, fine-bioclastic, and planktonic-micritic facies; it is interpreted as debris derived from rudist-reef deposits.

Fine-bioclastic facies (Fig. 16A, B)—Light-gray to tan, fine- to medium-grained skeletal limestone with sparry-calcite cement makes up this facies. The fossils consist principally of rounded, moderately well-sorted fragments of rudists; gastropods, corals, and foraminifers also are important constituents of the skeletal part of

the facies. Steeply dipping laminae, stylolites, and vertical and horizontal fractures are common. Much of the fine-bioclastic facies is recrystallized, leached, and dolomitized. Porosity is high in some cores, particularly in those from the P. Alemán No. 94 and Escolín No. 186, where production is in part from the fine-bioclastic facies.

In and east of the New Golden Lane, the fine-bioclastic facies is present in cores from two wells. In the E. Ordóñez No. 5 it is interbedded with the coarse-bioclastic facies and at one interval forms part of the producing zone. In the Cazones No. 2 the fine-bioclastic facies is recorded from strata of Jurassic age near the bottom of the well, where the fragments consist of solenoporid algae, pelecypods, and smaller foraminifers. The facies is interpreted as a shallow-shelf shoal-water deposit.

In the Poza Rica trend, the fine-bioclastic facies is present in all but one of the wells from which cores were studied. It is interbedded with the coarse-bioclastic, detrital, planktonic-micritic, and dolomite facies. Recrystallization and dolomitization of the fine-bioclastic facies locally are intense, as much as 100 percent. The porosity is high in the altered and fractured limestone. A pseudobrecciated texture is evident in several highly recrystallized cores of this facies. The fine-bioclastic facies in the Poza Rica trend is interpreted as skeletal debris derived from rudist reefs; it is interbedded locally with rudist-reef deposits and with deeper water, open-marine limestones.

Detrital facies (Fig. 16C, D)—Light-gray, tan, and dark-brown detrital limestone consisting of angular to subangular clasts of consolidated rock of the coarse-bioclastic, fine-bioclastic, and planktonic-micritic facies in a micritic-skeletal or micritic matrix characterizes this facies. Stylolites and fractures are common; pyrite is sparsely distributed. Planktonic foraminifers are sparse to abundant.

Laminae within individual clasts are oriented at high angles to laminae within other clasts in the same rock. The matrix generally lacks evident bedding. Thin graded beds and convolute beds are uncommon.

In most examples of the detrital facies, the clasts are primarily of bioclastic limestone. In a few wells, however, and especially in the Zapotalillo No. 9, clasts of planktonic-micrite are embedded in a micritic matrix. The detrital facies is interpreted to have formed by a mixing of lithologies from different depositional environments through a process of rock fall and submarine slumping, in conjunction with faulting or possibly small-scale turbidity flows down the steep slopes

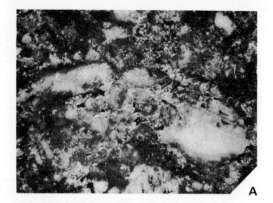

Fig. 15—Coarse bioclastic facies of Golden Lane and Poza Rica trend.

A. Coarse-grained skeletal limestone, leached, and oil stained. From depth of 2,955 m (9,675 ft) in San Andrés No. 143. Ground surface, ×1.5.

B. Coarse-grained skeletal limestone with sparry calcite cement. From depth of 2,279 m (7,475 ft) in Zapatalillo No. 9. Negative print of peel, ×3.

C. Coarse-bioclastic facies. Coarse-grained gastropod-skeletal limestone with sparry calcite cement and partial micritic-skeletal matrix. From depth of 2,611 m (8,565 ft) in P. Alemán No. 94. Negative print of peel, ×3.

off rudist reefs in the Poza Rica trend. The facies is interbedded with coarse-bioclastic facies, fine-bioclastic facies, and planktonic-micritic facies; it is not porous, except where fractured.

Planktonic-micritic facies (Fig. 17)—Dark- to light-gray, brown, and black micritic, micritic-skeletal, and argillaceous micritic limestone characterizes the planktonic-micritic facies. Small-scale irregular laminae, burrows, and pyrite are common. The limestone is replaced locally by chert and dolomite in many cores from wells in the Poza Rica trend. Vertical and horizontal fractures and stylolites are common. The fauna comprises common to abundant planktonic foraminifers, tintinnids, radiolarians, and planktonic crinoids; sparse mollusks and ostracods; and, in one core, miliolid foraminifers.

Intergranular porosity is very low to lacking, but permeability may be relatively high in rocks having abundant fractures. In the Jiliapa No. 61 a few meters of highly dolomitic limestone of this facies is part of the producing interval.

The planktonic-micritic facies is one of the most widespread facies, present in all but two of the Poza Rica trend wells studied and in the Tuxpan No. 2 and Cazones No. 2 east of the Golden Lane. In the Tuxpan No. 2 this facies was found in cores of the Agua Nueva Formation, directly overlying the El Abra Limestone. In the Cazones No. 2 it was present in two cores of Valanginian age (earliest Cretaceous) at a depth equivalent to similar facies of the same age in the lower Tamaulipas Formation, just west in the Poza Rica trend. In the Poza Rica wells the planktonic-micritic facies ranges in age from Late Jurassic to Late Cretaceous (Cenomanian). It is interpreted as a quiet-water, open-shelf, or basinal deposit.

Pseudodetrital facies—Light-gray, highly recrystallized limestone and dolomitic limestone of the pseudodetrital facies have megascopic textures and structures that appear similar to those of the detrital facies but, on close inspection, do not have a matrix of planktonic-bearing micrite or a mixture of lithologies from different environments. The pseudodetrital facies is recognized only in cores which could not be assigned to other, better formulated facies based on primary rock textures.

The "pseudobreccia" described by Barnetche and Illing (1956) from the lower Tamabra Limestone in the Poza Rica field corresponds to the pseudodetrital facies. Barnetche and Illing (1956, p. 16) stated that "a brecciated appearance is common in the cores of the Lower Tamabra Limestone of the Poza Rica No. 86 and Poza Rica No. 95 due to 'fragments' about one centi-

meter across, largely recrystallized so as to destroy most of their clastic structure, and left as subrounded patches of very finely crystalline groundmass of lime sands and lime silts. Commonly, these patches have been selectively dolomitized suggesting that they are a pseudobreccia." The use of "pseudodetrital" in this report differs from the use of "pseudobreccia" in that pseudodetrital is used here only for rock which is so strongly altered that it could not be related to a more specific carbonate facies. Much of the core assigned to the coarse-bioclastic facies also has a pseudobrecciated or pseudodetrital aspect, especially if small core segments are viewed individually.

Porosity in the pseudodetrital facies locally is high, where it is apparently the result of alteration of bioclastic limestone. The producing interval in the P. Aleman No. 94 is partly in this facies.

ENVIRONMENTAL INTERPRETATIONS

Alternatives

The outstanding production history of the Golden Lane and the Poza Rica trend has interested many geologists whose papers have resulted in several conflicting interpretations of the depositional environment and structural history of the reservoir trends. Two interpretations are reviewed here to illustrate the notable differences.

Interpretation 1

A widely accepted interpretation (Fig. 18A), modified in detail many times in 30 years, depicts the El Abra Limestone along the Golden Lane as a reef, especially a barrier reef or, more recently, as an atoll-edge reef, and deposits on the Golden Lane platform as lagoonal (Muir, 1936; Rockwell and García-Rojas, 1953; Guzmán, 1967; Viniegra and Castillo-T., 1968, 1970). The Tamabra Limestone of the Poza Rica trend is interpreted as forereef and turbidity deposits derived from the El Abra Limestone reef and interbedded with basinal sediments (Bonet, 1963; Boyd, 1963; Becerra, 1970). The Tamaulipas Limestone is considered a fine-grained basinal deposit. This interpretation requires a substantial part of the 1,000-m (3,280 ft) relief between the Golden Lane and the Poza Rica trend to have been developed during the Early and middle Cretaceous (Fig. 6).

Interpretation 2

Another interpretation (Fig. 18B), somewhat modified from Barnetche and Illing (1956), suggests that the El Abra Limestone represents a shelf deposit; the Tamabra Limestone a mixture

Fig. 16—Facies of Poza Rica trend.

A. Fine-bioclastic facies. Steeply dipping beds composed of skeletal limestone with vuggy porosity and uneven oil stain. From depth of 2,598 m (8,520 ft) in P. Alemán No. 94. Ground surface, ×1.5.

B. Fine-bioclastic facies. Fine-grained skeletal limestone with abundant intergranular pore space (black). From depth of 2,695 m (8,840 ft) in Escolin No. 186. Negative print of thin section, ×10.

C. Detrital facies. Large bioclastic and steeply dipping micritic-skeletal limestone fragments. From depth of 2,232 m (7,310 ft) in Zapotalillo No. 9. Negative print of peel, ×3.

D. Detrital facies. Subangular clasts of planktonic-micritic and bioclastic limestone mixed in micritic matrix. From depth of 1,892 m (6,205 ft) in Soledad No. 124. Ground surface, ×1.5.

E. Detrital facies. Large clast of rudist skeletal limestone in dipping, laminated micritic limestone. From depth of 2,230 m (7,305 ft) in Zapotalillo No. 9. Negative print of peel, ×3.

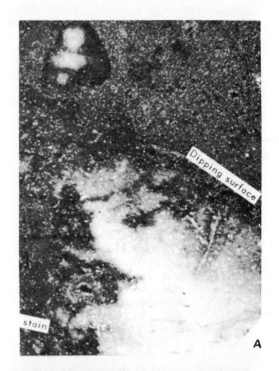

A

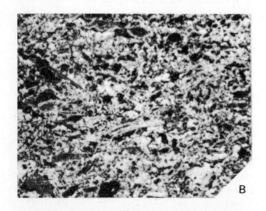

B

C

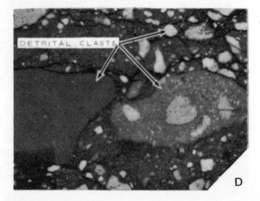

D

E

of in-place reef and locally derived forereef deposits, and the Tamaulipas Limestone a basinal or open-shelf deposit. In this interpretation, the 1,000-m (3,280 ft) difference in elevation between the Golden Lane and Poza Rica trend would be attributed to faulting or other postdepositional movement, and the narrow trend of the Golden Lane production attributed to the tilted edge of the Golden Lane platform.

Interpretation Resulting from This Study

Previous environmental interpretations of the Golden Lane and Poza Rica trend, with the exception of Barnetche and Illing (1956) and Becerra (1970), are based on ideas derived mainly from little or no core study; instead, they combine early observations of samples, stratigraphic and structural relations, and seismic data. The environmental interpretations offered here result from an intensive study of lithology, textures, structures, and faunal distribution based on examination of vertically sawed cores and thin sections. These data were fitted into the regional framework available from Pemex district offices at the time of the loan of the cores and from subsequently published work.

The El Abra Limestone in the New Golden Lane and in the center of the Golden Lane platform represents a shallow-shelf or lagoonal deposit. The presence of anhydrite (Tuxpan No. 2) suggests that supratidal conditions may have existed locally on the platform. A few rudist bodies, analogous to the well-known outcropping rudist bodies of the Fredericksburg Group in central Texas, are present in the New Golden Lane wells. East of the New Golden Lane, older middle Cretaceous limestone was deposited on a shallow shelf, and in the Cazones No. 2 there is a record of the development of oolite banks and associated sediments. Below the oolite banks, the earlier Cretaceous consists of limestone deposited in deeper water (lower Tamaulipas Formation). No core was studied from the Old Golden Lane; apparently none remains to be studied.

Interestingly, the original interpretation of the El Abra Limestone as reef was made by Muir (1936) from samples blown out of Old Golden Lane wells. Muir's descriptions were similar to ours for cores from the El Abra Limestone in the E. Ordóñez No. 5. The latter are not considered to be reefal here, and Viniegra and Castillo-T. (1970) agreed with this interpretation. On the basis of Muir's scanty and no longer available data, we would compare the Old Golden Lane closely with the E. Ordóñez No. 5 environmentally. Viniegra and Castillo-T. (1970), faced with the same facts but convinced of the reefal nature

of the Golden Lane, hypothesized that the reef has been eroded away (their Fig. 4) to leave erosional remnants consisting only of the lagoonal facies which now trap the oil. In fact, no one to our knowledge has ever published descriptions of coral-rudist reef from wells of the Golden Lane fields, Old, New, or Marine. Paul Enos (SUNY, Binghamton, New York), who briefly looked at cores of the New Golden Lane wells Mesita No. 1 and Cañas No. 101 mentioned by Viniegra and Castillo-T. (1970), reported that these make compelling analogs of the Taninul and El Abra facies of the Sierra del Abra (Bonet, 1952; Rose, 1963). We do agree with Viniegra and Castillo-T. (1970) "that little or nothing is known about the genesis of the Golden Lane reef, partly because information is sparse or incomplete, but mainly because no detailed analytic studies have been made to summarize the information from all of the wells, most of which have penetrated only a few meters into the top of the reef structures." We have described the core from two Golden Lane wells and from two wells on the platform, and have searched for all other written, intelligible descriptions. All the extant lithologic data from Golden Lane wells known to us lack any evidence of coral-rudist reef in the Golden Lane.

Many environments are represented in the cores of the Tamabra and upper Tamaulipas Formations of the Poza Rica trend. The Tamabra Limestone is composed largely of shallow-water, reef-derived debris. In addition, several intervals are interpreted as undisturbed, in-place rudist and coral-rudist reef. The fossil occurrences (Fig. 19) and sequence of facies in the Tamabra Limestone (for example, in the San Andrés No. 143), do not support the theory that the Tamabra originated as a slide and turbidite deposit derived from an El Abra atoll-edge reef. Two examples may suffice.

First, the core of the San Andrés No. 143 shows a logical, expectable, upward-shoaling sequence of carbonate facies (Fig. 10) beginning with the planktonic-micritic facies at the bottom of the core and progressing upward through coarse-bioclastic, fine-bioclastic, caprinid-micritic to coarse-bioclastic limestones. Such a sequence can be interpreted readily as open marine, reef front, reef core, and backreef. It is difficult to imagine that such a sequence, 90 m (295 ft) thick, could have been transported 6 km (3.7 mi) as a discrete block onto a structural high. It is even more difficult to imagine the sequence developing piecemeal. Secondly, miliolid foraminifers, which are extremely abundant in the El Abra Limestone, are very scarce in the many Tamabra cores

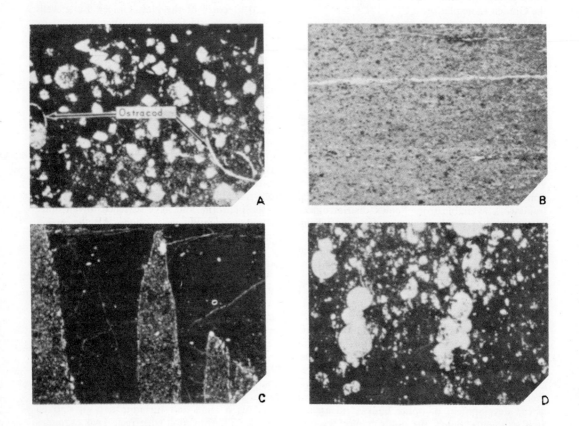

Fig. 17—Planktonic-micritic facies of Poza Rica trend.

A. Dolomitic, planktonic-micritic limestone. Ostracods and planktonic microfossils are abundant. From depth of 3,040 m (9,970 ft) in San Andrés No. 143. Thin section, ×40.

B. Laminated micritic limestone with abundant planktonic organisms. From depth of 2,312 m (7,583 ft) in San Lorenzo No. 2. Negative print of peel, ×3.

C. Fractured and stylolitic micritic limestone. From depth of 2,708 m (8,882 ft) in Escolin No. 186. Negative print of peel, ×3.

D. Globigerinid-micritic limestone. From depth of 1,854 m (6,080 ft) in Soledad No. 124. Thin section, ×40.

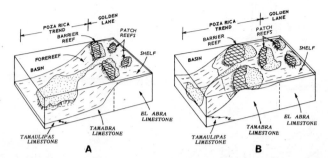

F<small>IG</small>. 18—Diagrammatic reconstructions of two alternative geologic settings for environment of deposition of middle Cretaceous formations in Golden Lane and Poza Rica trend. **A.** Interpretation 1, Tamabra Limestone solely as forereef, slope, and basinal deposits. **B.** Interpretation 2, Tamabra Limestone as reef and associated deposits.

studied. Conversely, the reef organisms in the Tamabra Limestone (coral, radiolitid rudists, stromatoporid hydrozoans, encrusting algae, boring pelecypods, and larger foraminifers, especially abundant specimens of *Dictyoconus*) were not found in the four El Abra Limestone wells studied.

Interbedded with reef debris of the Tamabra Limestone are units of planktonic-foraminiferal micritic limestone which were deposited in deeper, open-marine water. Where the micritic limestone lacks reef-derived debris, it is called the upper Tamaulipas Limestone. The detrital Tamabra Limestone which formed from clasts of lithified rudist-reef mixed with planktonic foraminiferal micritic limestone probably represents reef talus (Fig. 16C); it is common in wells near the edges of producing fields. The fact that the bioclastic debris comprising some of the clasts was cemented prior to deposition as part of detrital limestone may be evidence of some Cretaceous subsea cementation.

Stratigraphic and Structural Information

An isopach map (Fig. 20) of the Upper Jurassic and the Lower Cretaceous (El Abra-Tamabra-

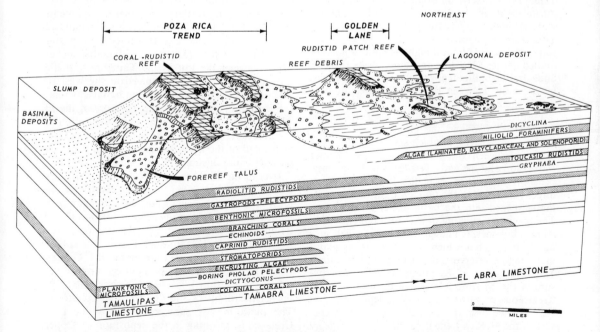

F<small>IG</small>. 19—Diagrammatic reconstruction of environment of deposition of El Abra, Tamabra, and Tamaulipas Limestones for small area based on this study, together with distribution of important fossil groups.

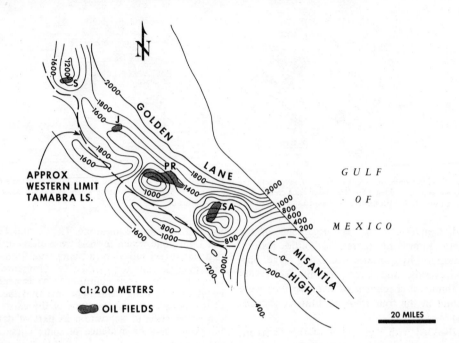

FIG. 20—Isopach map of Upper Jurassic and Cretaceous to top of the El Abra, Tamabra, and Tamaulipas Limestones in southern part of Golden Lane and Poza Rica trends. S = Soledad field, J = Jilipa field, Pr = Poza Rica field, SA = San Andrés field. (Modified from unpublished Pemex map.)

Tamaulipas Limestones) shows localization of Tamabra reef facies on isopach thins and of producing fields in the Poza Rica trend coincident with, or slightly offset from, these isopach thins. This relation suggests that the Tamabra coral-rudist reefs grew on local highs of either folded or faulted earlier Mesozoic and older rocks (Fig. 7). Slide and turbidite deposits would have accumulated mainly between these reef-topped highs—e.g., in the Zapotalillo No. 9.

Further evidence for the relation between the Poza Rica trend and the Golden Lane is provided in cross sections which show that there is a breccia between the Late Cretaceous Méndez Formation and the early Tertiary basal Chicontepec Formation (Fig. 7) in the Poza Rica trend, directly adjacent to the Golden Lane and extending westward into the basin about 8 km (5 mi). The breccia, consisting in part of fragments of El Abra lithologies, is known from wells which produce from it, but which were seeking as a primary target reservoir the Tamabra Limestone. The breccia, therefore, offers evidence of exposure of the El Abra Limestone during the latest Cretaceous or earliest Tertiary independently of the evidence from the Golden Lane wells. Exposure of the Golden Lane is confirmed by Viniegra and Castillo-T. (1970) who cite a limestone cavern filled with marine Oligocene sediments in the Mozutla No. 1, New Golden Lane. The Tamabra Limestone is buried under Late Cretaceous and early Tertiary open-marine sediments and could not have been exposed subaerially at the end of the Cretaceous. It is possible to postulate at least 100 m (328 ft) of differential movement between the top of the Tamabra and the top of the El Abra Limestones on the western side of the Golden Lane platform by latest Cretaceous time to account for the exposure of the El Abra Limestone and the origin and deposition of the breccia (Fig. 7). Eustatic lowering of sea level at the end of the Mesozoic also may have been a primary cause of subaerial exposure.

If both the El Abra and Tamabra Limestones were deposited at or very near sea level, then we must look for evidence of postdepositional movement which resulted in the present 1,000-m (3,280 ft) difference in elevation between the top of the two formations (Fig. 6). Such evidence is shown on the longitudinal section along the Golden Lane (Fig. 5). There is a pronounced arch to the top of the El Abra Limestone which is caused by the eastward tilt of the Golden Lane platform. The platform was submerged during the Late Cretaceous, resulting in the deposition of Upper Cretaceous open-marine sediments (e.g., Agua Nueva Formation and others) over the platform

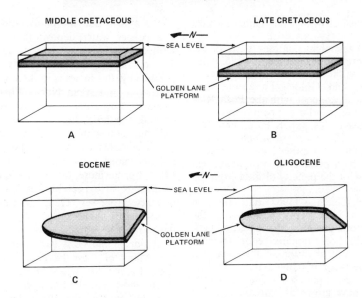

Fɪɢ. 21—Diagrammatic reconstruction from west to east, showing relative position of Golden Lane platform and top of El Abra Limestone to sea level as result of post-middle Cretaceous structural movements (not to scale). Diagrams do not consider possible effects of eustatic changes in sea level. **A.** Middle Cretaceous (Albian-Cenomanian)—top of El Abra Limestone near sea level. **B.** Late Cretaceous—platform sinks and receives Late Cretaceous sediments (Agua Nueva, San Felipe, Méndez Formations). **C.** Eocene—platform tilts downward to west allowing deposition of thick Eocene sediments on west-central platform edge. Model implies possible exposure of eastern edge depending on eustatic sea level changes. **D.** Oligocene—platform tilts downward to east allowing thick Oligocene sediments to be deposited on northern and southern flanks of western edge of platform and implies possible exposure of western edge.

(Viniegra and Castillo-T., 1970). After exposure in the latest Cretaceous or early Tertiary, the platform was tilted down toward the west (Fig. 21) allowing for the accumulation of a thick section of Eocene clastic deposits over the Old Golden Lane (Fig. 5). After the Eocene, tilting again must have been toward the east, resulting in the present arch to the top of the El Abra Limestone. The Eocene strata are structurally high and thick at the center of the Old Golden Lane arc, and the Oligocene and Miocene sections are thick and low north and south of the center of the arch. Tilting was suggested in part by Sotomayor-C. (1954) to explain the relations.

Dɪsᴄᴜssɪᴏɴ

In substance, the interpretation of the Tamabra as shoal-water, mixed reef-complex and the Golden Lane primarily as lagoon is more comprehensive but similar in concept to the ideas offered by Barnetche and Illing (1956) and disputed by many geologists since then. This interpretation of the origin of the rocks of the Golden Lane and Poza Rica trend is known already to many Pemex geologists and others interested in the petroleum geology of Mexico (Guzmán, 1967). The differences in interpretation are not trivial for the origin of the Tamabra Limestone,

nor for the interpretation of the cause of 1,000 m (3,280 ft) of structural relief separating the Poza Rica trend from the Golden Lane. The differences are less important for the Golden Lane, for it is no longer necessary to consider the Golden Lane as a barrier reef in order to explain its oval trend, reservoir characteristics, trapping mechanisms, or platform-edge production.

Remarkably, either interpretation of the Golden Lane appears in retrospect to be an adequate basis for exploration of the onshore and Marine Golden Lane, probably because the trapping of hydrocarbons in structural highs along the Golden Lane platform edge could be independent of the origin of the highs, whether the result of reefal growth structures or erosion and tilting. Certainly the reef interpretation of the Golden Lane has served well as an exploration philosophy.

Neither interpretation explains the lack of productive Tamabra reservoirs west of most of the Old Golden Lane fields or in the yet unexplored area east of the Marine Golden Lane. The reef interpretation would have to account for a lack of Tamabra turbidite deposits originating from the Old Golden Lane along the structurally highest area; the Tamabra reef-complex interpretation would have to account for a lack of middle

Cretaceous "highs" on which the Tamabra reefs could grow.

Other alternatives might combine parts of the two basic interpretations. Barnetche (1963), for example, shows a "line of reefs" which extends south along the Old Golden Lane and then swings westward to connect with the Poza Rica trend reefs; however, he did not comment on the basis for the idea in his paper. Parenthetically, having a line of reefs crossing a structural trend seems less plausible than either basic interpretation, but would be a possible environmental explanation in view of the paucity of data from the Old Golden Lane.

More recently, Guzmán (1967) has taken an intermediate view. He accepts the idea that the Tamabra Limestone contains *in-situ* rudist reefs, citing our unpublished company report (Bebout *et al.*, 1965), without abandoning the idea that the Golden Lane is a reef-fringed atoll. He does this by considering the possibility that the Tambra reefs might have grown as individual groups paralleling the Golden Lane atoll. The lack of absolute proof or disproof of a Golden Lane platform-edge reef leaves this as a viable interpretation if it does not discount structural movement to separate the Golden Lane and Poza Rica trend after the middle Cretaceous.

Criticism of our interpretation of the El Abra of the Golden Lane as primarily lagoonal with scattered rudist bodies of the bank and patch reef variety (Fig. 19) and the Poza Rica trend as true reef fronting on a western deeper water basin with reef development principally on old "highs" is based on the following arguments:

1. The relief on the Golden Lane scarp must be attributed to post-Tamabra movement, and there is no detailed structural rationale or mechanism to account for the manner or form of this movement.

Although we are not primarily structural geologists, it is clear from the facies and stratigraphy of the Tampico embayment that subsidence was the dominant movement in both the Golden Lane and Poza Rica areas after the middle Cretaceous, causing the deposition of the deep-water Agua Nueva, San Felipe, and Méndez Formations over both areas. In fact, the history of the Early and middle Cretaceous also is predominantly one of subsidence. Exposure of the El Abra Limestone along the edge of the Golden Lane platform in the early Tertiary requires uplift and possibly lower sea level, and may be accounted for by tilting similar to that shown in Figure 21. Tilting presumably involves faulting.

2. The abundance of pelagic (planktonic-micritic facies) units in the Poza Rica trend indicates deep-water deposition of the Tamabra Limestone.

It is agreed that the planktonic-micritic facies is a deeper water deposit than the others described. It is important to note also the distribution of these planktonic-micritic units. Stratigraphically, this facies is in the upper Tamaulipas Limestone and Agua Nueva Formation; in the Tamabra Limestone it is interbedded with the detrital, fine-bioclastic, and coarse-bioclastic facies but not with the caprinid-micritic facies, which contains the rudist bodies interpreted as in-place rudist reefs (Figs. 10, 11; Bebout *et al.*, 1965). Furthermore, in our opinion, the sequence of units is an expectable result of normal sedimentary processes. Geographically, the planktonic-micritic facies units, which commonly are thin, are more numerous on the western or basinal side of the Poza Rica trend, except at the top. Many of these have been mapped in the Poza Rica field, where Barnetche and Illing (1956) show that the units pinch out toward the east. Of course, the Agua Nueva Formation which covers the Tamabra Limestone appears to encroach gradually over the Poza Rica trend (Figs. 7, 11), as shown in the Escolín No. 186, Jiliapa No. 61, and Soledad No. 124 wells. Cores from wells such as the Zapotalillo No. 9, located off the old "highs," may show a dominance of slope and basinal deposits.

3. The great thickness of the El Abra Limestone in the Golden Lane platform and the presence of rudist bodies in this position can be accounted for only through a reefal origin of the Golden Lane.

Comparison with the Great Bahama Banks, which also are constructional carbonate platforms having more than 4,600 m (15,000 ft) of shallow-water carbonate rock, shows clearly that carbonate platforms need not have significant amount of reef contribution either to maintain their positions near sea level or to continue growth in the face of continued subsidence.

4. The degree of fragmentation of the rudist debris in the coarse bioclastic facies is evidence that these particles were transported a long distance from the Golden Lane to a bathyal site of final deposition.

Study of modern carbonate sedimentation shows that the most abraded and broken shells are in the intertidal zone where wave attack is fierce and constant, or where organic reworking is intense; for example, on beaches and bars or in lagoons and on reefs. There is comparatively little wear during downslope movement into bathyal depths, as in shown in cores from the Tongue of the Ocean.

5. The poor preservation of bioclastic fragments and the lack of orientation of the valves of mollusks are evidences of transportation.

Poor preservation caused by constant abrasion or organic reworking is certainly characteristic of bioclastic particles in the intertidal and shallow subtidal zone, and such abraded and coated particles are found easily today in Florida and the Bahamas. Chemical solution is more prevalent in the intertidal and supratidal zones, where carbonate shells are subjected to water chemically different from the sea in which they grew. The quiet marine waters of the bathyal zone (the 1,000-m-deep "Tamabra sea"?) should have provided an environment relatively free of abrasion and solution for rapidly deposited and buried shells, if above the calcium carbonate compensation zone.

It is difficult to tell whether rudists are correctly oriented, even if one is a specialist on rudists; it is very difficult in cores. However, the orientation of caprinid rudists in a few beds of the Tamabra Limestone is quite similar to the orientation of those from undisputed, in-place, organic-bound rudist banks in Texas.

6. The presence of argillaceous sediments in typical Tamabra Limestone reinforces the evidence of the pelagic origin of the Tamabra rocks, because argillaceous sedimentation is incompatible with the development of rudist banks.

First, it should be noted that in the E. Ordóñez No. 5 argillaceous sediments also were deposited as part of the El Abra Limestone. Second, the Agua Nueva Formation is argillaceous and directly overlies the El Abra Limestone on the Golden Lane platform. Finally, outcrop studies in Texas show that interbedding of argillaceous carbonate rocks with nonargillaceous rudist limestones of bank or patch-reef origin, within a few feet vertically, can be seen where the Walnut and Comanche Peak Formations change vertically and laterally into the Edwards Limestone. Argillaceous content in itself is obviously no argument for deep-water deposition.

SIGNIFICANCE FOR EXPLORATION

Regardless of the correctness of the reef hypothesis for the origin of the Golden Lane, it now appears that the bulk of evidence calls for reservoir development through leaching of erosional highs developed by subaerial karst formation during the Tertiary (Viniegra and Castillo-T., 1970). The platform edge, especially on the western side, probably was weathered most severely because of tilting and sea level drop. Hence, the strategy of searching for petroleum in structural highs on the platform edge has been so successful.

Presumably the interpretation of the Golden Lane as a barrier reef or atoll-edge reef evokes an exploration strategy in which one looks for Tamabra reservoirs in turbidite and slump deposits around the whole periphery on the Golden Lane platform. The interpretation of the Tamabra as coalescing reef complexes developed on basement highs calls for exploration within a much smaller area adjacent to the platform. If the folding or faulting below the Poza Rica trend (Figs. 7, 20) did indeed localize Tamabra reef growth, and if these structures are related either to the Misantla high or the Sierra Madre Oriental, then it would not be expected that another Poza Rica trend would be found seaward of the Marine Golden Lane, except possibly along its southern edge.

COMPARISON OF GOLDEN LANE AND POZA RICA TRENDS WITH "DEEP EDWARDS" TREND OF SOUTH TEXAS

Comparisons of the "Deep Edwards" trend with the Golden Lane have been attempted by many geologists (including Boyd, 1963; Rose, 1963) interested in Cretaceous reservoirs in south Texas. Tempting analogies with the Golden Lane "reef" trend and its porosity characterize most discussions. The following summary of similarities and differences is directed largely to a comparison of the facies, environments of deposition, and geologic histories of the two areas. Pertinent references in south Texas literature include Keith (1963) and Tucker (1962).

Similarities

1. Shallow-water facies of the Golden Lane and Poza Rica trend and of the "Deep Edwards" trend developed adjacent to a deeper, open-water facies.

2. Faulting on the basinward side is indicated for both the Golden Lane and the "Deep Edwards" trend.

3. The carbonate rocks in both areas are approximately the same middle Cretaceous age.

4. The rocks overlying the El Abra and Tamabra Limestones and those overlying the Edwards Limestone are open shelf or basinal in many areas and are similar in sequence, lithology, and fauna. These overlying deposits indicate that, after deposition of the shallow-shelf and reef facies, both areas were covered next by deeper water deposits.

5. Formations in the Golden Lane, Poza Rica trend, and "Deep Edwards" trend are predominantly carbonate rocks with rudist pelecypods as an important constituent.

6. Several facies of the Golden Lane and Poza

Rica trend are similar to those of the "Deep Edwards" trend.

Differences

1. The Golden Lane is 50-65 km (30-40 mi) from the Sierra Madre Oriental, a major early Tertiary (Laramide) orogenic belt. This orogeny probably caused tilting of the Golden Lane platform and exposure of part of the El Abra Limestone to subaerial weathering which developed considerable vuggy and cavernous porosity. The "Deep Edwards" trend is hundreds of kilometers from the Laramide belt and was not exposed to surface weathering after the middle Cretaceous.

2. The equivalent of the Poza Rica trend in structural and geographic position and degree of postdepositional alteration has not been found along the "Deep Edwards" trend of south Texas.

3. The fauna and flora of the Golden Lane differ from those of the "Deep Edwards" trend because different biofacies are involved: (a) the Golden Lane facies and faunas compare more closely with the backreef and shelf equivalents of the Stuart City reef than with the rudist-reef facies of the "Deep Edwards" trend; and (b) hydrozoans and algae, which are important reef binders in the Edwards Limestone Stuart City reef, e.g., in the Tenneco No. 1 Schulz (Keith, 1963), were found only in wells of the Poza Rica trend in Mexico.

4. Shallow-shelf molluscan marl and shale similar to the Walnut and Comanche Peak Formations of Texas have not been noted to any extent from the Tampico embayment.

5. The Golden Lane and Poza Rica trend are prolific oil-producing areas; the Edwards Limestone-Stuart City reef trend is only marginally productive.

In summary, the depositional facies of the Golden Lane and Poza Rica trend are broadly similar to those of the Edwards Limestone and equivalent formations of the "Deep Edwards" trend and the backreef lagoon in south Texas, but the subsequent geologic histories of the two regions are markedly different. The differences in the postdepositional history affected the reservoir characteristics, the type of traps, and the movement of fluids. In eastern Mexico these features resulted in the development of giant oil reservoirs, but no reservoirs of great economic importance have been found in the "Deep Edwards" trend of south Texas.

References Cited

Acosta-Estévez, R., 1968, Petroleum developments in Mexico in 1967: Am. Assoc. Petroleum Geologists Bull., v. 52, p. 1354-1365.

——— 1969, Petroleum developments in Mexico in 1968: Am. Assoc. Petroleum Geologists Bull., v. 53, p. 1564-1577.

——— 1970, Petroleum developments in Mexico in 1969: Am. Assoc. Petroleum Geologists Bull., v. 54, p. 1587-1597.

Acuña-G., A., 1957, El distrito petrolero de Poza Rica: Asoc. Mexicana Geólogos Petroleros Bol., v. 9, p. 505-553.

Barnetche, A., 1963, Lobos Island—Mexico's most challenging discovery: Petroleum Management, v. 35, no. 13, p. 67-73.

——— and L. V. Illing, 1956, The Tamabra Limestone of the Poza Rica oil field, Veracruz, Mexico: 20th Internat. Geol. Cong., Mexico, 38 p.

Bebout, D. G., A. H. Coogan, and C. M. Maggio, 1965, Facies of the Golden Lane and the Poza Rica trend: Humble Oil and Refining Co., Exploration Dept., Geol. Research Sec., Rept. 65-3, 27 p., unpub.

——— 1969, Golden Lane-Poza Rica trends, Mexico—an alternate interpretation (abs.): Am. Assoc. Petroleum Geologists Bull., v. 53, p. 706.

Becerra H., A., 1970, Estudio bioestratigráfico de la Formación Tamabra del Cretácico en el distrito de Poza Rica: Inst. Mexicana Petróleo Rev., v. 2, no. 3, p. 21-25.

Bonet, F., 1952, La facies Urgoniana del Cretácico medio de la región de Tampico: Asoc. Mexicana Geólogos Petroleros Bol., v. 4, p. 153-262.

——— 1963, Biostratigraphic notes on the Cretaceous of eastern Mexico, in Geology of Peregrina Canyon and Sierra de El Abra, Mexico: Corpus Christi Geol. Soc. Ann. Field Trip Guidebook, p. 36-48.

Boyd, D. R., 1963, Geology of the Golden Lane trend and related fields of the Tampico embayment, in Geology of Peregrina Canyon and Sierra de El Abra, Mexico: Corpus Christi Geol. Soc. Ann. Field Trip Guidebook, p. 49-56.

Díaz-González, T., 1969, Exploration and production results of offshore extension of Mexican Cretaceous Golden Lane (abs.): Am. Assoc. Petroleum Geologists Bull., v. 53, p. 715.

Franco, A., 1969, La actividad marina se agiganta: Petróleo Interamericano, v. 27, no. 9, p. 25-36.

Figueroa-H., S., 1964, Nueva a los 55 es la Faja de Oro: Petróleo Interamericano, v. 22, no. 8, p. 56-58.

Guzmán, E. J., 1967, Reef type stratigraphic traps in Mexico, in Origin of oil, geology and geophysics: London, Elsevier, 7th World Petroleum Cong., Mexico, Proc., v. 2, p. 461-470.

Keith, J. W., 1963, Environmental interpretation of subsurface Washita-Fredericksburg limestones, northern Live Oak County, south Texas, in Geology of Peregrina Canyon and Sierra de El Abra, Mexico: Corpus Christi Geol. Soc. Ann. Field Trip Guidebook, p. 72-78.

López-Ramos, E., 1959, Origen del petróleo en relación a las cuencas de depósito: Asoc. Mexicana Geólogos Petroleros Bol., v. 11, p. 155-167.

Mina-Uhink, F., 1966, Petroleum developments in Mexico in 1965: Am. Assoc. Petroleum Geologists Bull., v. 50, p. 1553-1563.

——— 1967, Petroleum developments in Mexico in 1966: Am. Assoc. Petroleum Geologists Bull., v. 51, p. 1435-1444.

Muir, J. M., 1936, Geology of the Tampico region, Mexico: Am. Assoc. Petroleum Geologists, 280 p.

Murray, G. E., and P. R. Krutak, 1963, Regional geology of northeastern Mexico, in Geology of Peregrina Canyon and Sierra de El Abra, Mexico: Corpus Christi Geol. Soc. Ann. Field Trip Guidebook, p. 1-10.

Rockwell, D. W., and A. García-Rojas, 1953, Coordination of seismic and geologic data in Poza Rica-Golden Lane area, Mexico: Am. Assoc. Petroleum Geologists Bull., v. 37, p. 2551-2565.

Rose, P. R., 1963, Comparison of the type El Abra of Mexico with "Edwards Reef trend" of south-central Texas, in Geology of Peregrina Canyon and Sierra de El Abra, Mexico: Corpus Christi Geol. Soc. Ann. Field Trip Guidebook, p. 57-64.

Sotomayor-Castañeda, A., 1954, Distribución y causas de la

porosidad en las calizas del Cretácio medio en la región de Tampico, Poza Rica: Asoc. Mexicana Geólogos Petroleros Bol., v. 6, p. 157-206.

Tucker, D. R., 1962, Subsurface Lower Cretaceous stratigraphy, central Texas, *in* Contributions to geology, south Texas: South Texas Geol. Soc., p. 177-216.

Viniegra, F., and C. Castillo-Tejero, 1968, Golden Lane fields, Veracruz, Mexico (abs.): Am. Assoc. Petroleum Geologists Bull., v. 52, p. 553-554.

———— ———— 1970, Golden Lane fields, Veracruz, Mexico, p. 309-325, *in* Geology of giant petroleum fields: Am. Assoc. Petroleum Geologists Mem. 14, 575 p.

World Oil, 1969, v. 169, no. 3, p. 72-73.

The American Association of Petroleum Geologists Bulletin
V. 59, No. 4 (April 1975), P. 665-693, 25 Figs.

Secondary Carbonate Porosity as Related to Early Tertiary Depositional Facies, Zelten Field, Libya[1]

D. G. BEBOUT[2] and CHARLES PENDEXTER[3]

Houston, Texas 77001

Abstract Production from the Zelten field, Libya, is from the highly porous shelf limestones of the Zelten Member ("Main Pay") of the Paleocene and lower Eocene Ruaga Limestone. Fifteen facies are recognized, mapped, and predicted. Seven of the facies comprise the larger part of the Zelten Member. These include miliolid-foraminiferal micrite, argillaceous bryozoan/echinoid micrite, argillaceous-molluscan micrite, coralgal micrite, *Discocyclina*-foraminiferal calcarenite, foraminiferal calcarenite and micrite, and *Discocyclina*-foraminiferal micrite.

In the Zelten field secondary porosity is recorded as much as 40 percent; this porosity is related to the original depositional fabric of the sediment and is, therefore, facies controlled. Porosity is highest in the coralgal micrite and *Discocyclina*-foraminiferal calcarenite, which together form a northwest-southeast trend across the northern part of the field, and in the foraminiferal calcarenite and micrite. The rocks of these three facies primarily are grain supported with a micrite matrix in which the primary carbonate mud porosity was preserved because of the lack of compaction. Subsequent leaching by ground water through these porous zones probably enlarged the primary mud porosity and greatly altered the original carbonate texture. Porosity is lowest in the miliolid-foraminiferal-micrite and argillaceous bryozoan/echinoid micrite facies, both of which are blanketlike in distribution over the top of the field and are the caprock for the reservoir. Porosity also is low in the argillaceous molluscan-micritic facies that forms a lens-shaped body in the southern part of the field southwest of the coralgal-micrite and *Discocyclina*-foraminiferal-calcarenite facies.

A thick dolomite in the lower part of the Zelten Member is restricted to the area south of the trend formed by the coralgal micrite and *Discocyclina*-foraminiferal calcarenite. Thus, these sediments probably were deposited in a lagoonal area more susceptible to dolomitization. The coincidence of the dolomite and its off-structure position, however, leaves open the possibility for structural control.

Introduction

Although there are large quantities of oil in Tertiary carbonate sequences, little has been published concerning the nature of reservoir-trap depositional facies, particularly in nonreefal sections. A notable exception is the excellent paper by Terry and Williams (1969), in which they described the Idris "A" bioherm in the Sirte basin. Even less is known about the relation of the carbonate facies to the exceptional porosities developed by secondary processes. Knowing facies are porous is not enough, however, unless other occurrences or extensions of these facies can be predicted; this can be done by relating the porous facies to depositional environments whose distributions are predictable.

Geologic Setting

The Zelten field is approximately 100 mi south of the Mediterranean coast in the center of the Cretaceous-Tertiary Sirte basin of Libya (Fig. 1). The Zelten Member ("Main Pay") of the Ruaga Formation and underlying Heira Shale are considered by Esso Libya geologists to be Paleocene; the overlying member, the Meghil, is dated as early Eocene (Fig. 2).

In 1966, 56 Zelten wells produced 159,000,000 bbl of 40° API oil from the Zelten Member, for an average of more than 7,700 bbl/day/well (Bowerman, 1967, p. 1565); in 1972, 187 wells produced 84,627,825 bbl (Nicod, 1973). Cumulative production for the field exceeded 1.5 billion barrels during 1972 (Nicod, 1973).

Production is from the highly porous Zelten Member, and porosity is both of the primary intergranular and intragranular and secondary leached types. The argillaceous upper part of the Zelten Member and the Meghil Member form the caprock for the reservoir.

Cretaceous block faulting divided the Sirte basin into a series of northwest-southeast trending ridges and troughs (Fig. 3). The Late Cretaceous transgression on these exposed blocks resulted in the development of quartz and carbonate shoreline sands that form the reservoirs of many Cretaceous fields of the Sirte basin. Contemporaneous with these shoreline sands, deeper water silt with planktonic organisms was deposited in the

[1]Manuscript received, May 31, 1974; accepted, August 1, 1974.

[2]Exxon Production Research Company; presently with Bureau of Economic Geology, The University of Texas at Austin.

[3]Exxon Production Research Company.

Cooperation of many Esso Libya geologists, particularly W. V. Naylor, H. A. Jarvis, R. B. King, and L. A. Smith, is gratefully acknowledged. Exxon Production Research Company personnel consulted during this study include Stefan Gartner, Jr. (coccoliths), N. K. Brown, Jr. (foraminifers), D. G. Harris (porosity types), and J. E. Faris (reservoir engineering). R. M. Mitchum, Jr., assisted in logging the core and in correlating the electrical and radioactivity logs.

Appreciation is extended to Exxon Production Research Company and Esso Libya for granting permission to publish the results of this study.

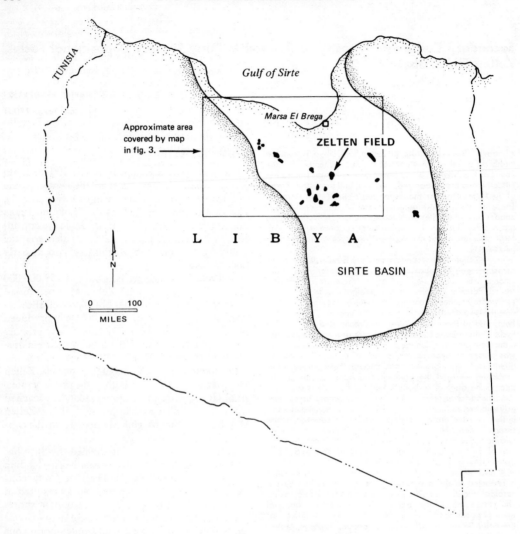

FIG. 1—Location of Sirte basin and Zelten field.

troughs. In the Paleocene, shallow-water carbonate sediments within the Heira Shale (Mabruk, Ora, and Meem) accumulated on some of the ridges, such as the Zelten ridge, until late Paleocene and early Eocene (Zelten and Meghil Members of the Ruaga), when carbonate sedimentation spread over the entire basin (Fig. 4). The preexisting Cretaceous structures are still evident in the Tertiary, however, because shallow-water carbonate materials were deposited on the buried highs, and deeper water carbonate materials in the troughs. The transgression, which began in the Late Cretaceous, seemingly ended in the early Paleocene. The following regressive sequence culminated within the Zelten Member, and the overlying Meghil Member was deposited at the beginning of the next major transgression.

FACIES ANALYSES AND ENVIRONMENTAL INTERPRETATIONS

Facies types identified from detailed descriptions of cores and cuttings (see Figs. 7-13) have been correlated by means of a series of cross sections. The correlation grid into which these sections were placed was based mostly on electric and radioactivity logs. The most reliable detailed correlations within such a thin unit as the Zelten Member, however, are difficult to make unless the lateral facies variations are known. Only the facies section B-B' (Figs. 5, 6) is included in this paper;

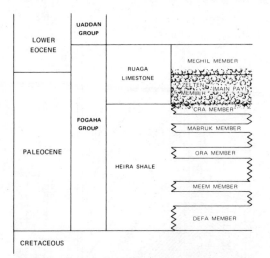

FIG. 2—Lower Tertiary stratigraphic section Sirte basin. Paleocene and lower Eocene Ruaga Limestone consists of seven members—Meghil, Zelten, Cra, Mabruk, Ora, Meem, and Defa. Basinal shale equivalent of lower five limestone units is Heira Shale. Stratigraphic nomenclature presented is entirely Esso Libya usage.

it is presented on a stratigraphic datum corresponding to the top of the Meghil Member, and porosity and permeability charts have been superimposed on the facies sections to demonstrate better the relation of porosity to facies (Fig. 6).

Interpretation of the original carbonate texture and fossil content in many of the Zelten cores is difficult because of the high degree of recrystallization, alteration, leaching, and oil staining. Failure to account for these postdepositional changes can result in establishment of diagenetic facies rather than depositional facies. Fossils, for instance, have a definite order of recrystallization or selectively are destroyed because of differences in the original composition of the shell, the size, shape, and orientation of crystals that make up the shell, the proportion of foreign ions, and the permeability of the matrix material (Banner and Wood, 1964).

A total of 15 facies was recognized in this study; the most common and important facies are as follows.

Miliolid-foraminiferal micrite (Fig. 7) and argillaceous bryozoan/echinoid micrite (Fig. 8)—The mi-

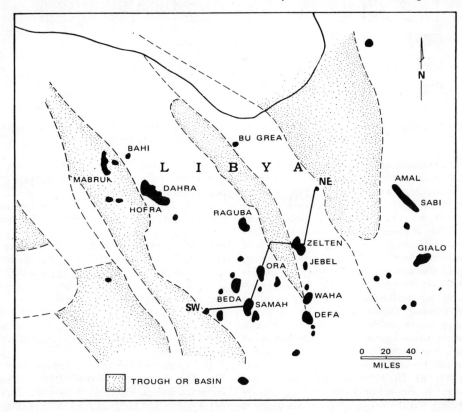

FIG. 3—Upper Cretaceous structural setting of Sirte basin. Figure 1 shows location of map. Cross section shown on Figure 4.

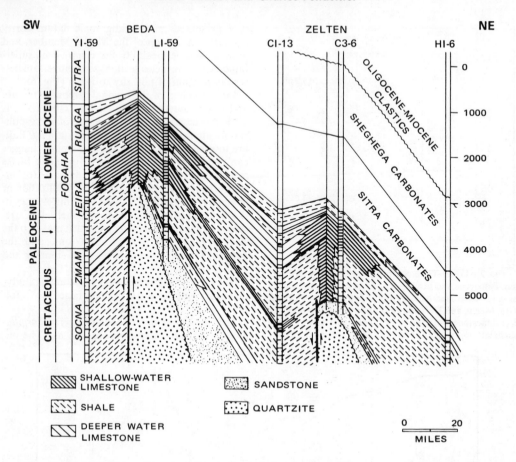

FIG. 4—Regional cross section across Sirte basin along line shown on Figure 3. Section crosses two highs, indicated by position of quartzite and intervening trough.

liolid-foraminiferal micrite and argillaceous bryozoan/echinoid micrite are essentially argillaceous micritic limestones with burrows and irregular laminae. Miliolids, bryozoans, and echinoids are the dominant fossils. These facies are in the upper part of the Zelten Member and, because of their blanketlike distribution (Fig. 6) and low porosities and permeabilities, form the caprock for the reservoir.

Argillaceous molluscan micrite (Fig. 9)—The argillaceous molluscan-micrite facies consists of argillaceous micritic limestone with burrows and irregular laminae. The limestone is similar to the preceding two facies, but it contains an abundant molluscan fauna. This facies is in the central and southern part of the field, south of the coralgal-micrite and *Discocyclina*-foraminiferal-calcarenite facies (Fig. 6). The porosity of the argillaceous molluscan micrite is rather low (5-10 percent, as high as 30 percent locally), resulting in a decrease in porosity from north (area of coralgal micrite and *Discocyclina*-foraminiferal calcarenite) to

south in the field. Porosity data from the southern part of the field are not reliable because of the poor quality of porosity logs.

Coralgal micrite (Fig. 10)—The coralgal-micrite facies comprises a loose framework of branching and laminated corals with a micrite matrix. This facies is present in two zones in the northern part of the field—in the lower Zelten, laterally equivalent to the dolomite, and in the upper Zelten. The coralgal micrite commonly is limited in distribution to a circular area 4-5 mi in diameter (Fig. 6) and thins rapidly southward. Effective porosity is typically high (30-35 percent) in this facies, the result of highly altered and leached micrite and leached corals.

Discocyclina-foraminiferal calcarenite (Fig. 11)—The *Discocyclina*-foraminiferal calcarenite facies is a skeletal limestone composed of foraminiferal grains that either are very poorly cemented with spar or have a micrite matrix. This facies is present in several thick pods which, together with

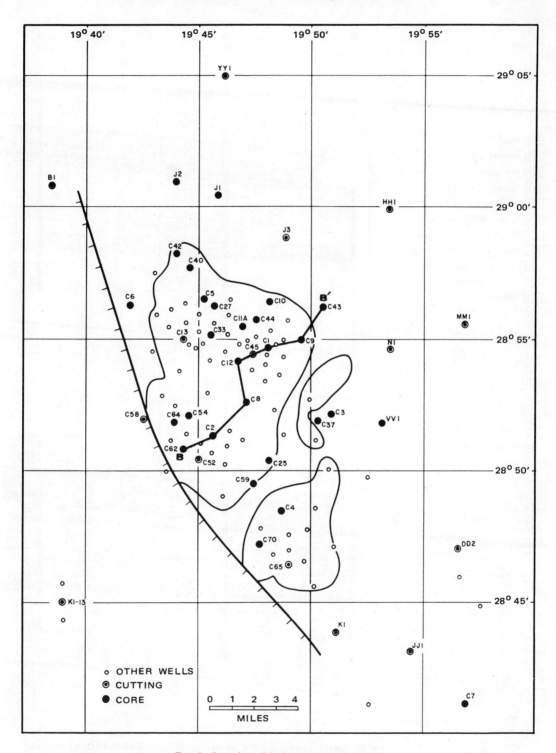

FIG. 5—Location of facies cross section *B-B'*.

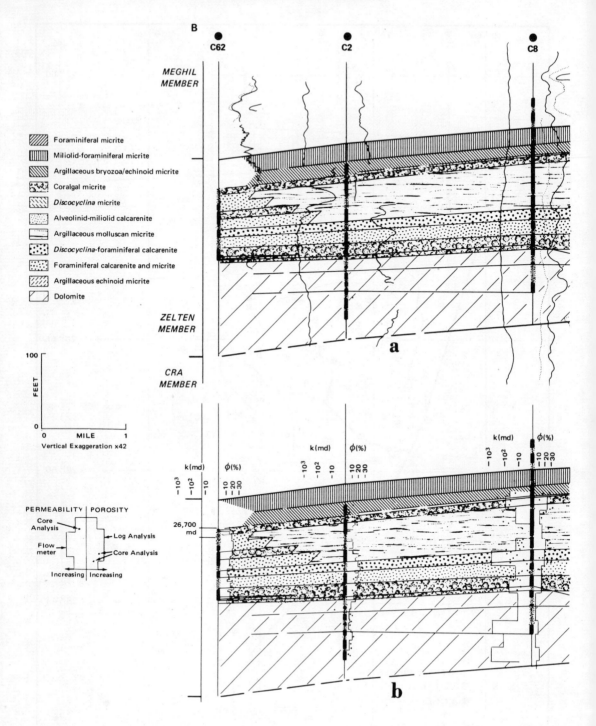

Foraminiferal micrite

Miliolid-foraminiferal micrite

Argillaceous bryozoa/echinoid micrite

Coralgal micrite

Discocyclina micrite

Alveolinid-miliolid calcarenite

Argillaceous molluscan micrite

Discocyclina-foraminiferal calcarenite

Foraminiferal calcarenite and micrite

Argillaceous echinoid micrite

Dolomite

Vertical Exaggeration x42

PERMEABILITY | POROSITY

Core Analysis

Log Analysis

Flow meter

Core Analysis

Increasing | Increasing

FIG. 6—Facies cross section *B-B'* on stratigraphic datum (top of Meghil Member)

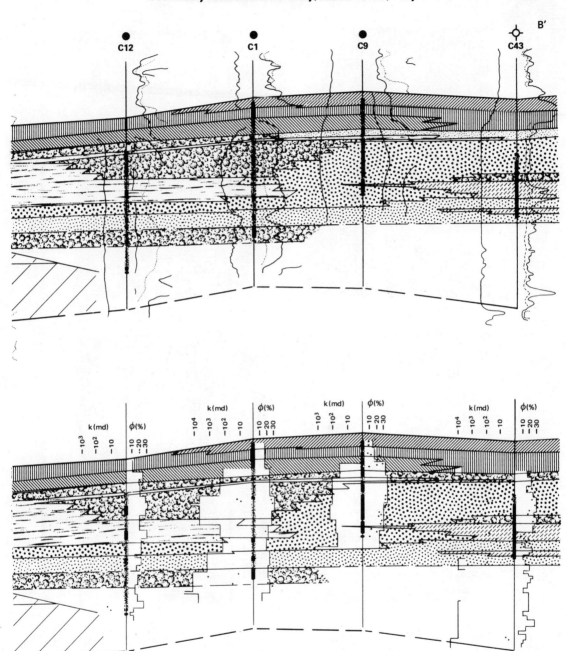

showing (a) facies analysis and (b) porosity and permeability. See Figure 5 for location.

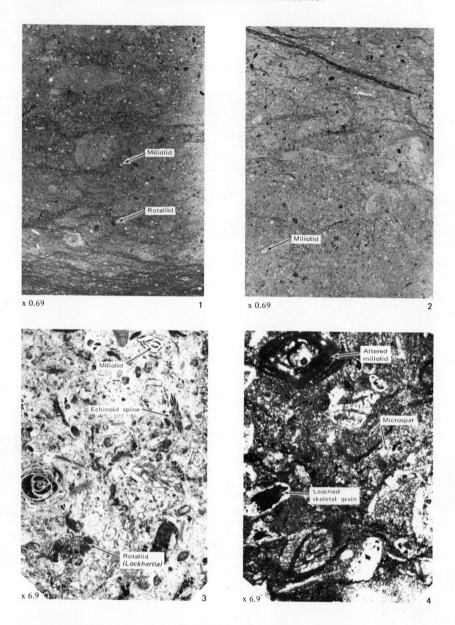

FIG. 7—Miliolid-foraminiferal micrite. Lithology: limestone to argillaceous limestone; up to 20 percent argillaceous material. Texture: dominantly micritic-skeletal limestone, locally skeletal-micritic; 20-80 percent skeletal, 10-80 percent micrite. Structures: abundant burrows and irregular laminae. Minerals: pyrite, glauconite locally. Recrystallization: none to 70 percent (very thin zones). Porosity: leached skeletal and micrite; none to fair where altered. Dominant fossils (in this and other figure captions: A, abundant; C, common; R, rare; O, lacking): miliolids (R-A), rotaliids (C-A, mostly species of *Lockhartia*), *Discocyclina* (O-R), planktonic foraminifers (R), alveolinids (O-C); other fossils are arenaceous foraminifers (R-A), *Operculinoides* (O-A), articulated and crustose coralline algae (O-C), dasycladacean algae (C) very locally, *Distichophax* (C) locally, echinoids (R-A), bryozoans (O-A), pelecypods (A), gastropods (A), *Gryphaea* (O-A) very locally, worm tubes (O-C).

1. C8-6, depth 5,710 ft. Large rotaliids (*Lockhartia*) are dark and miliolids are light. This contrast is typical in this facies. 2. C37-6, depth 5,933 ft. 3. C10-6, depth 5,799 ft. Typical unaltered tight micrite with abundant miliolids and other skeletal debris. 4. C37-6, depth 5,933 ft. Porosity is developed locally, as shown by alteration of micrite matrix to microspar and leaching of skeletal grains.

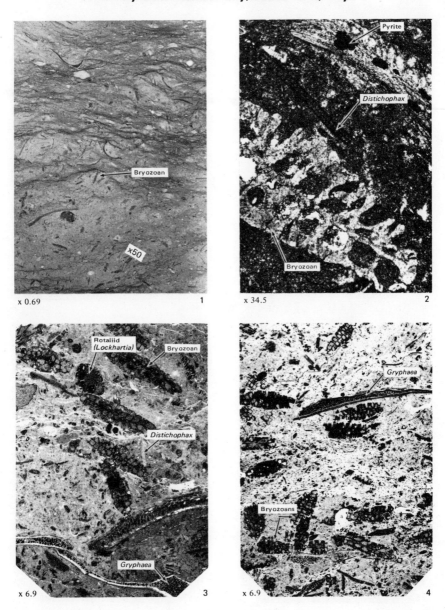

FIG. 8—Argillaceous bryozoan echinoid micrite. Lithology: argillaceous limestone; up to 20 percent argillaceous material. Texture: dominantly micritic-skeletal limestone; 30-80 percent skeletal, 20-70 percent micrite. Structures: abundant burrows and irregular laminae. Minerals: scattered chert and pyrite. Recrystallization: varies from 0-90 percent, most commonly very low. Porosity: leached skeletal and micrite; commonly very low quality, locally good, particularly in lower part. Dominant fossils: bryozoans (O-A), echinoids (O-A), miliolids (O-A), planktonic foraminifers (R) locally, alveolinids (O-R), *Distichophax* (O-C); others are arenaceous foraminifers (R-C), *Operculinoides* (R-C), *Discocyclina* (R), rotaliids, dasycladacean algae (R) locally, crustose coralline algae (O-A) in abundance with *Gryphaea* and worm tubes, horn, branching, and massive colonial corals (A) in thin zones, pelecypods (C-A), *Gryphaea* (R-A), gastropods (C-A), worm tubes (O-A).

1. C3-6, depth 5,882 ft. Argillaceous laminae are common. 2. C1-6, depth 5,460 ft. Porosity is commonly low in this unaltered micrite; some intragranular porosity, however, is preserved within bryozoan. 3. C1-6, depth about 5,460 ft. 4. C33-6, depth 5,628 ft. Argillaceous material is light colored on this negative print. Bryozoans are typically most abundant where argillaceous content is highest.

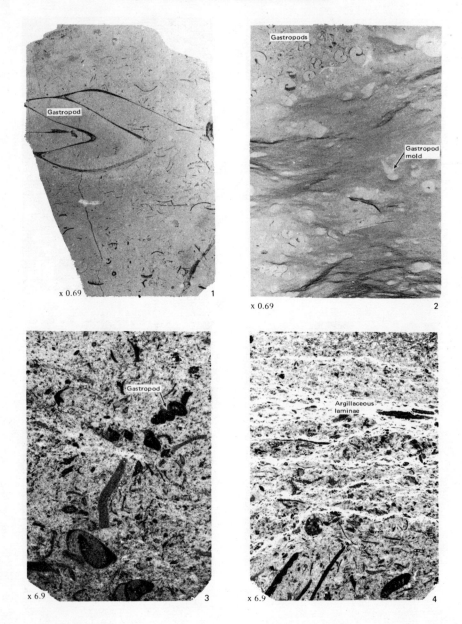

FIG. 9— Argillaceous molluscan micrite. Lithology: argillaceous limestone; argillaceous content up to 20 percent; scattered dolomite rhombs 0-10 percent. Texture: typically micritic-skeletal limestone; 20-70 percent skeletal, 30-80 percent micrite. Structures: none. Minerals: scattered chert and pyrite. Recrystallization: ranges from 0-100 percent altered, most commonly less than 50 percent altered. Porosity: leached micrite and skeletal; poor quality where not altered, good where highly altered. Dominant fossils: gastropods (C-A), pelecypods (C-A); others are arenaceous foraminifers (O-C), miliolids (O-C), alveolinids (O-C), *Operculinoides* (R), *Discocyclina* (R-C), rotaliids (R-C), crustose coralline algae (O-C), dasycladacean algae (R-C), bryozoans (R-C), *Gryphaea* (C-A), worm tubes (O-C), echinoids (R-A).

1. C40-6, depth 6,020 ft. 2. C3-6, depth 5,910 ft; abundant argillaceous laminae; internal molds of gastropods appear similar to burrows. 3. C3-6, depth 5,949 ft. 4. C2-6, depth 5,820 ft; abundant argillaceous laminae shown in white.

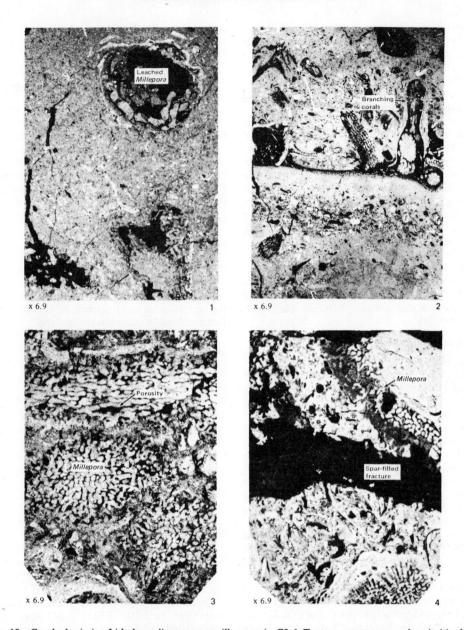

FIG. 10—Coralgal micrite. Lithology: limestone, argillaceous in C8-6. Texture: most commonly micritic-skeletal limestone; 30-70 percent skeletal, 10-70 percent micrite, 0-20 percent spar cement. Structures: few irregular laminae, common parallel laminae formed by laminated *Millepora*; brecciation of laminae common. Minerals: pyrite associated with corals; highly mineralized zone in lower coralgal unit of C44 contains pyrite, marcasite, pyrrhotite, bornite, strontianite, chalcopyrite, galena, chalcocite, and sphalerite. Recrystallization: 0-100 percent; commonly very highly altered. Porosity: leached micrite and skeletal (corals); large vugs in lower coralgal zone; quality, typically very good, fair to poor where not altered. Dominant fossils: articulated and crustose coralline algae (C-A), branching corals (C-A), *Millepora* (C-A), boring pelecypods (O-C); others are arenaceous foraminifers (C-A), miliolids (O-A)—abundant in lower coralgal zone—*Discocyclina* (O-R), *Operculinoides* (C), rotaliids (C-A), alveolinids (C) locally, dasycladacean algae (O-C), echinoids (C), massive colonial corals (O-A), horn corals (O-A), pelecypods (O-A), *Gryphaea* (O-R), gastropods (O-C).
     1. C1-6, depth 5,636 ft; leached *Millepora* and very fine leached micrite porosity. **2.** C33-6, depth 5,670 ft. **3.** C1-6, depth 5,627 ft; very high leached skeletal porosity; dark areas are pores. **4.** C70-6, depth 5,884 ft.

the coralgal micrite, form an elongate trend from northwest to southeast across the northern part of the Zelten field and as a very thin zone through the center of the Zelten Member (Fig. 6). Porosity is typically high (30-35 percent) and of the primary intergranular and intragranular types that have been enhanced by leaching.

Foraminiferal calcarenite and micrite (Fig. 12)—A thick band of alternating foraminiferal calcarenite and foraminiferal micrite facies crosses the middle of the Zelten Member (Fig. 6). Because of the complex intermixture of these two facies, they are considered together. Porosity is commonly high (25-30 percent) in the northeast where the micritic zones are thickest; porous zones are thin and alternate with less porous zones (10-20 percent) in the southwest.

Argillaceous echinoid micrite and foraminiferal micrite (Fig. 13)—These facies comprise most of the section in the Zelten Member adjacent to the field. Both are dominantly micritic limestone with minor amounts of skeletal material. Porosity is commonly low.

INTERPRETATION OF VERTICAL SEQUENCE

The following interpretation of the vertical sequence is based on textural and faunal changes vertically through the section from the Heira Shale below to the top of the Ruaga Limestone (Meghil Member).

The Heira Shale is a basinal planktonic-foraminifer-bearing shale that grades upward into the Cra Member (basal member of the Ruaga). Typically, the Cra Member consists of planktonic-foraminifer micrite and *Discocyclina* micrite. This change indicates the beginning of a shallowing trend (regression) that can be traced upward in the Zelten Member. The basal Zelten facies is the *Discocyclina*-coralgal micrite, followed by coralgal micrite, foraminiferal calcarenite and micrite, and finally, *Discocyclina*-foraminiferal calcarenite and coralgal-micrite complex probably representing maximum regression.

A transgressive sequence begins with the argillaceous bryozoan/echinoid micrite and miliolid-foraminiferal micrite, the topmost facies of the Zelten Member. The *Discocyclina* micrite and planktonic micrite of the overlying Meghil Member complete the transgressive sequence.

Thus, the maximum regression occurred during deposition of the upper beds of the Zelten Member, when the carbonate deposits were subjected to subaerial exposure. The leached-micrite porosity and highly altered zones resulted from diagenesis of the carbonate matter by freshwater, a well-known mechanism for developing secondary porosity. The period of maximum regression corresponds very closely with the major regression at the close of the Paleocene in India, Pakistan, and Burma (Nagappa, 1959).

ENVIRONMENTAL INTERPRETATIONS

To interpret better the environment of deposition that each facies represents, a series of facies distribution maps has been constructed across seven horizons of the upper part of the Zelten Member (Figs. 14-21). These horizons are based on lithologic breaks interpreted from electric and radioactivity logs and on correlations of facies and fossils. The defining key beds or markers are considered time-parallel, separating isochronous units for analysis. Because of the poor control in many areas of the field, considerable interpretation was involved in preparing these maps.

In summary, this sequence begins with a broad shallow shoal covered by foraminiferal calcarenite and micrite (Fig. 15) that graded northeastward into offshore foraminiferal micrite. Later, during the regressive stage, calcarenite bars (*Discocyclina*-foraminiferal calcarenite) developed along a northwest-southeast trend and coralgal banks grew in passes between the bars (Figs. 16-18); lagoonal deposits accumulated behind (southwest of) this trend (argillaceous molluscan micrite). With deepening, the calcarenite bars no longer could persist and the sheetlike coralgal banks spread to the south. Offshore sediments (argillaceous bryozoan/echinoid micrite and foraminiferal micrite) then began to encroach over the area (Fig. 19). Additional deepening of the water finally resulted in covering the entire area with facies deposited in an offshore environment in which abundant planktonic foraminifers flourished (Figs. 20, 21).

POROSITY TYPES AND DISTRIBUTION

Porosity in the Zelten field ranges from less than 2 percent to more than 40 percent; porosity of more than 30 percent is common in the highly leached zones.

Five main types of porosity were recognized in cores from the Zelten field (Fig. 22). These can be summarized as follows.

Intragranular (Fig. 22e)—Primary porosity within skeletal grains that results from lack of filling of the body cavities of the organisms after death. This type can result in high but ineffective porosity unless combined with good intergranular or leached micrite porosity to connect the pores.

Intergranular (Fig. 22b, e)—Primary porosity that remains between the skeletal grains (particularly foraminifers and corals) or is trapped beneath them. Intergranular porosity accounts for some of the very high porosity in the Zelten field.

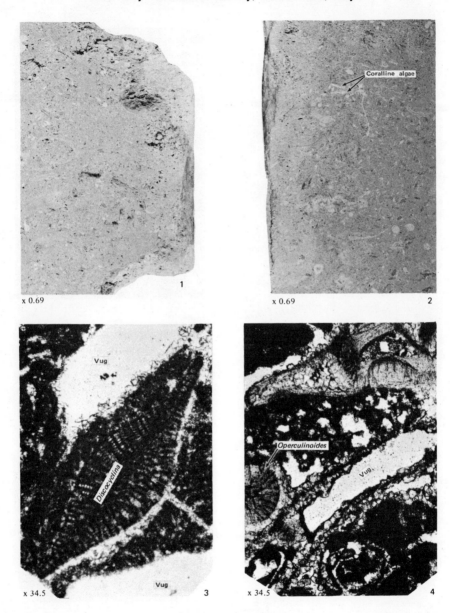

FIG. 11—*Discocyclina*-foraminiferal calcarenite. Lithology: limestone, up to 70 percent dolomite locally. Texture: skeletal-micritic limestone; 70-80 percent skeletal, 10-20 percent micrite, 10-20 percent spar cement. Structures: few scattered irregular laminae. Minerals: pyrite locally. Recrystallization: 0-100 percent altered, commonly highly altered and leached. Porosity: primary intragranular and intergranular, leached skeletal and micrite, very good quality where altered. Dominant fossils: *Discocyclina* (C-A), rotaliids (C-A), crustose coralline algae (O-A), articulated coralline algae (R-C); others are arenaceous foraminifers (O-R), miliolids (R-A), *Operculinoides* (R-A), alveolinids (O-R), *Nummulites* (O-C), dasycladacean algae (O-C), *Millepora* (O-A) locally, horn and branching corals (C-A) in scattered zones, echinoids (C-A), pelecypods (R-C), gastropods (R-A).

1. C45-6, depth 5,825 ft; highly altered and leached. 2. C9-6, depth 5,846 ft; highly altered and leached. 3. C43-6, depth 6,102 ft; vugs are probably result of leaching but may be in part primary intergranular. 4. C43-6, depth 6,107 ft; leached skeletal and leached micrite porosity.

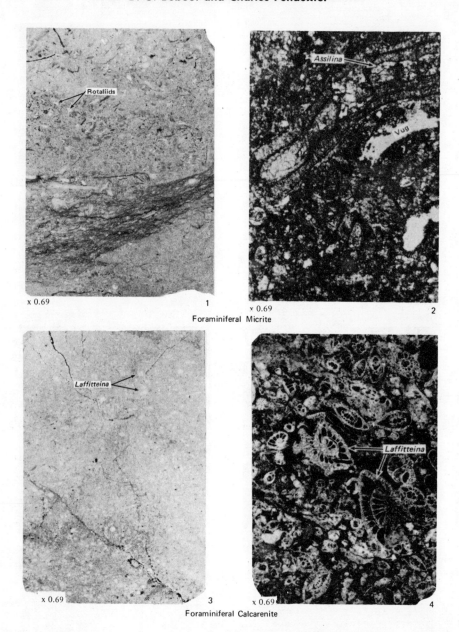

1 — x 0.69

2 — x 0.69

Foraminiferal Micrite

3 — x 0.69

4 — x 0.69

Foraminiferal Calcarenite

FIG. 12—Foraminiferal calcarenite and micrite. Lithology: limestone, argillaceous content up to 5 percent; dolomite in local zones from 20-80 percent. Texture: interbedded micritic-skeletal, skeletal-micritic, and skeletal limestones; 40-80 percent skeletal, 10-60 percent micrite, 10-20 percent spar cement. Structures: common irregular laminae, particularly in micritic zones. Minerals: none. Recrystallization: 0-100 percent; high in northern part of field; low and limited to thin zones in south. Porosity: leached micrite and skeletal, good quality where highly altered. Dominant fossils: rotaliids (C-A), *Laffitteina* (A) in lower part of facies, dasycladacean algae (R-A) in upper part of facies; others are arenaceous foraminifers (O-A), miliolids (O-C), *Operculinoides* (O-A), *Nummulites* (O-A) locally, articulated and crustose coralline algae (O-A), echinoids (O-A), pelecypods (O-A).

1. C54-6, depth 5,880 ft; porosity changes considerably in this single specimen; it is highly altered and porous at bottom and unaltered and tight at top. 2. C43-6, depth 6,157 ft; leached micrite porosity; micrite has been altered to microspar. 3. C43-6, depth 6,181 ft. 4. C33-6, depth 5,803 ft; good primary intergranular porosity; pores are lined with thin calcite druse.

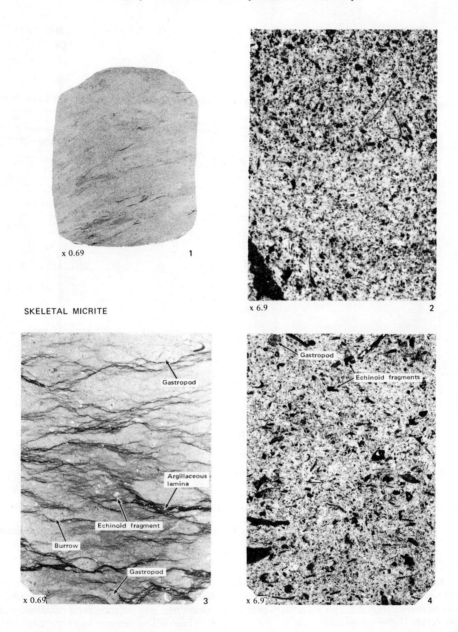

SKELETAL MICRITE

x 0.69   1

x 6.9   2

x 0.69   3

x 6.9   4

FIG. 13—Skeletal micrite. Lithology: limestone to argillaceous limestone. Up to 10 percent argillaceous material and 0-10 percent dolomite. Texture: most commonly skeletal-micritic limestone; 20-70 percent skeletal, 20-80 percent micrite. Structures: abundant irregular laminae and burrows. Minerals: none. Porosity: leached micrite, absent to poor. Dominant fossils: unidentifiable fine skeletal debris (A); others are arenaceous foraminifers (R), miliolids (O-A) and alveolinids (O-C) in only one zone, rotaliids (R-C), echinoids (C-A), pelecypods (R).

Argillaceous echinoid micrite. Lithology: argillaceous limestone with up to 10 percent argillaceous material. Texture: skeletal-micritic to micritic-skeletal limestone; 20-70 percent skeletal, 30-80 percent micrite. Structures: abundant irregular laminae and burrows. Vertical fractures in dolomite. Minerals: none. Recrystallization: none in limestone; 70-100 percent in dolomite. Porosity: vuggy dolomite, poor quality. Dominant fossils: echinoid fragments (A), others are arenaceous foraminifers (C), rotaliids (C-A), *Operculinoides* (O-C), pelecypods (O-C).

1. C25-6, depth 6,868 ft. 2. C25-6, depth 5,869 ft. 3. C3-6, depth 5,924 ft; abundant argillaceous laminae and burrows. 4. C8-6, depth 5,824 ft.

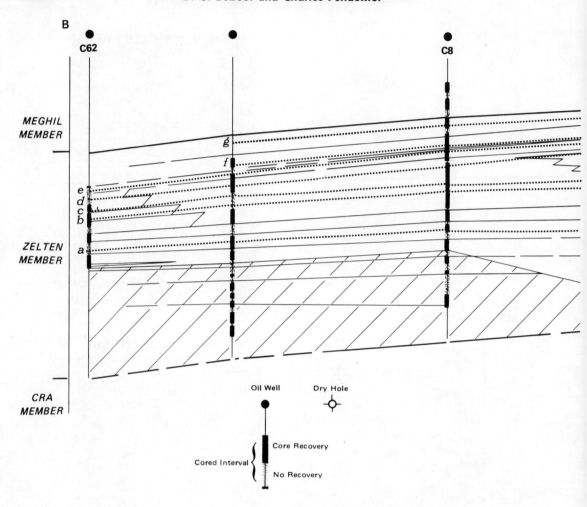

FIG. 14—Facies section *B-B'* showing location of facies horizon

Leached skeletal (Fig. 22d, f)—Secondary porosity that results from the solution of shell material.

Leached micrite (Fig. 22a, c)—Secondary porosity that results from recrystallization of the fine micrite particles to microspar, accompanied by leaching of carbonate. The fine pinpoint vug porosity that results accounts for much of the very high porosity in the Zelten field. Recrystallization as the mechanism for the formation of porosity is well known and has been described (Bathurst, 1958, 1959; Murray, 1960).

Dolomite vug—Secondary porosity formed during dolomitization. Very high porosity results, but this type of porosity is effective only if the vugs are connected by finer, intercrystalline porosity.

Porosity and permeability are best developed in the *Discocyclina*-foraminiferal-calcarenite, coralgal-micrite, and foraminiferal-calcarenite and micrite facies. The *Discocyclina*-foraminiferal calcarenite has both intergranular and intragranular porosity, modified by leaching and, in places, cementation. The coralgal micrite has both leached-skeletal (corals) and leached-micrite porosity. Both of these facies are lens-shaped in distribution (Figs. 23, 24) and together form a trend across the northern part of the Zelten field, where porosities and permeabilities are high and production is good and rather uncomplicated. In the south these porous facies thin abruptly, and the nonporous argillaceous molluscan micrite thick-

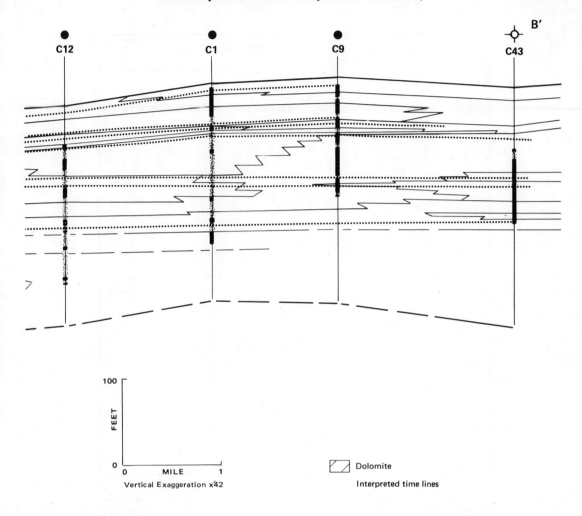

maps a-g (Figs. 15-21). See Figure 6 for location of cross section.

ens. The foraminiferal calcarenite and micrite have leached-skeletal, leached-micrite, and both intergranular and intragranular porosity. Porosity in this facies is high in the northern part of the field but, as a result of many thin interbeds of tight spar-cemented calcarenites, the porosity is less continuous vertically in the southern part. As far as can be determined, production in the southern part of the field is from the underlying dolomite.

Examination of sawed surfaces and thin sections indicates that the primary depositional fabric, especially the ratio of grains to mud, played an important role in later development of the secondary or leached porosity in the Zelten field. Porosity is best developed in three facies (*Discocyclina*-for-

aminiferal calcarenite, foraminiferal calcarenite and micrite, and coralgal micrite), all of which show evidence of being grain supported at an early stage. The calcarenite facies may have either carbonate mud or open spaces between the grains. Where carbonate mud is present, the original mud porosity, commonly more than 35 percent, is preserved because of the grain-supported nature of the sediment and lack of compaction. Thus, primary mud porosity provides avenues for later movement of ground water. Lucia (1962) found that the best porosity in Devonian crinoidal sedimentary rocks from Andrews County, Texas, is developed in those with 5-20 percent micrite and that porosity decreases as the micrite increases

205

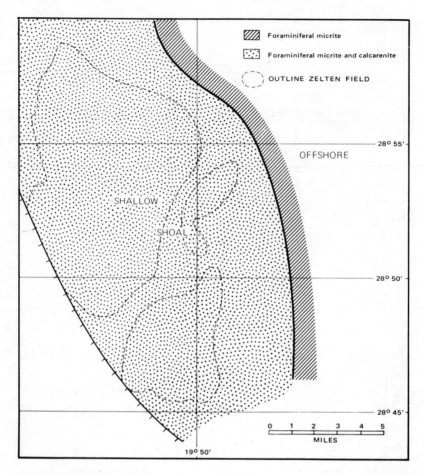

Fig. 15—Facies map of horizon a. A broad shoal area represented by foraminiferal calcarenite and micrite facies extends over entire Zelten field area. In east and northeast shoal-water facies changes abruptly to offshore foraminiferal micrite.

above 20 percent. As the percentage of micrite increases, the micrite itself must support the weight of the overlying sediments, and it is compacted, resulting in a loss of porosity. Lucia also found that the presence of mud in open spaces between grains inhibits the deposition of sparry calcite, and that the sediments with no micrite between grains tend to be plugged with spar.

Similarly, the high porosity of the coralgal micrite results from the lack of compaction of carbonate mud because of support afforded by the loose network or framework of branching and la-

minated corals and algae. Later solutions were able to pass through these porous muds, leaching them and the unstable aragonite corals.

DOLOMITE IN LOWER PART OF ZELTEN MEMBER

Dolomite in the lower part of the Zelten Member ranges in thickness from zero in the northern part of the field to more than 250 ft 5 mi south of the field (Fig. 25). The original limestone texture is difficult to identify because of the coarsely crystalline nature of the dolomite throughout much of the section. Several of the limestone facies (coral-

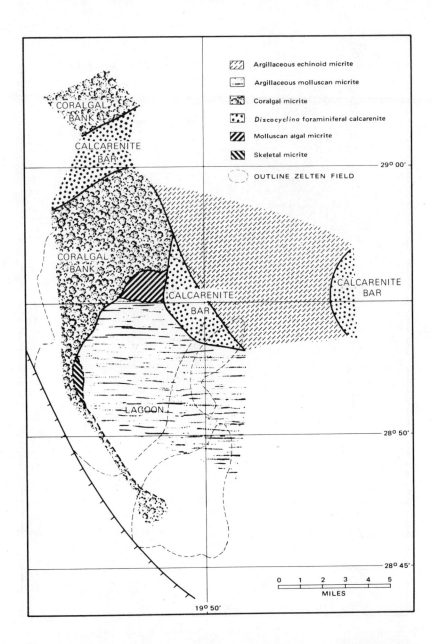

FIG. 16—Facies map of horizon b. Several small pods of *Discocyclina*-foraminiferal calcarenite are developing. Between calcarenite pods, which probably represent shallow-water bars, coralgal-micrite mud banks are growing. Elongate trend formed by combination of bars and mud banks restricts circulation in area in south sufficiently to allow deposition of lagoonal argillaceous molluscan micrite. Thin coralgal micrite zone south of the argillaceous molluscan micrite suggests that large coralgal body in northern part of Zelten field may have wrapped around western side of field.

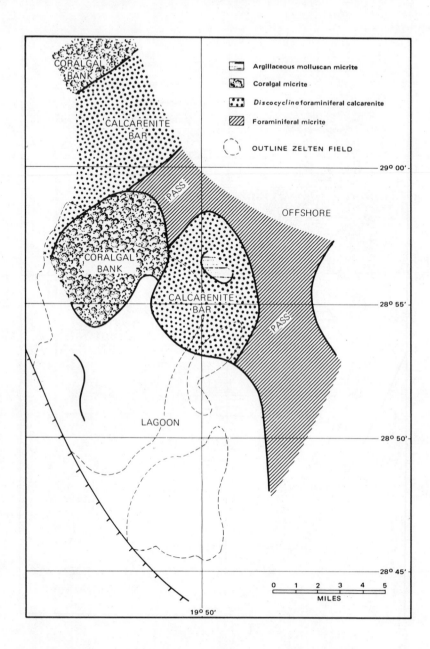

Fig. 17—Facies map of horizon c. Calcarenite bars are larger and better defined. Coralgal mud banks are more restricted and appear to be preferentially developed at head of pass (foraminiferal micrite) between bars. If this is true, another coralgal mud bank may be just southwest of easternmost calcarenite bar, where no control exists.

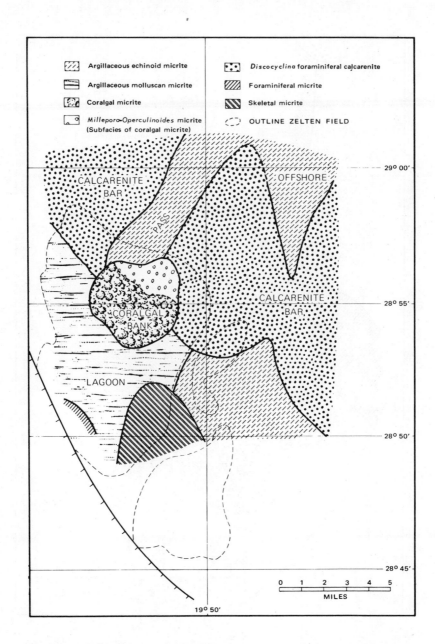

Fig. 18—Facies map of horizon d. Calcarenite bars are at maximum development. Pass between bars is narrow and coralgal mud bank again is at head of it. Quiet-water lagoon, in which argillaceous molluscan micrite was deposited, is still present behind calcarenite bars and coralgal mud banks.

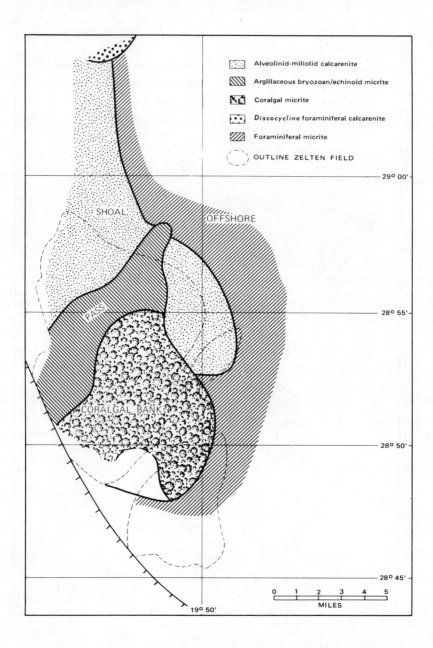

FIG. 19—Facies map of horizon e. With deepening of water calcarenite bars ceased to grow, but poorly developed shoal area is represented by alveolinid-miliolid calcarenite. Offshore foraminiferal micrite has moved in over much of area occupied previously by calcarenite. Pass now is represented by offshore argillaceous bryozoan/echinoid micrite facies. Coralgal bank, as result of deepening waters, has spread out southward as thin blanketlike body of sediment.

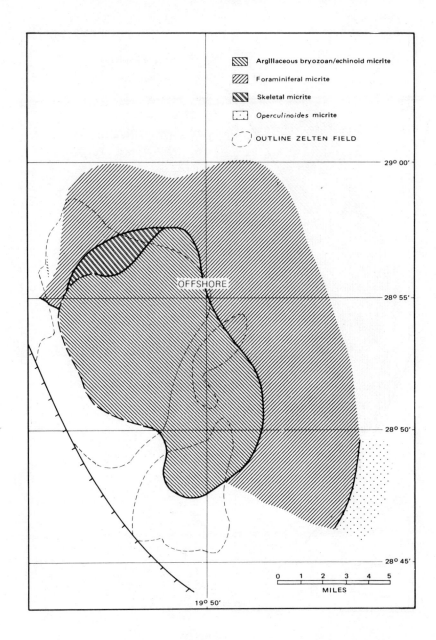

FIG. 20—Facies map of horizon f. With increasing water depth calcarenite bars and coralgal mud banks no longer can grow. Argillaceous bryozoan/echinoid micrite, seen only in pass on Figure 19, has spread over almost entire field area. Slightly deeper water around this argillaceous zone is represented by offshore foraminiferal micrite.

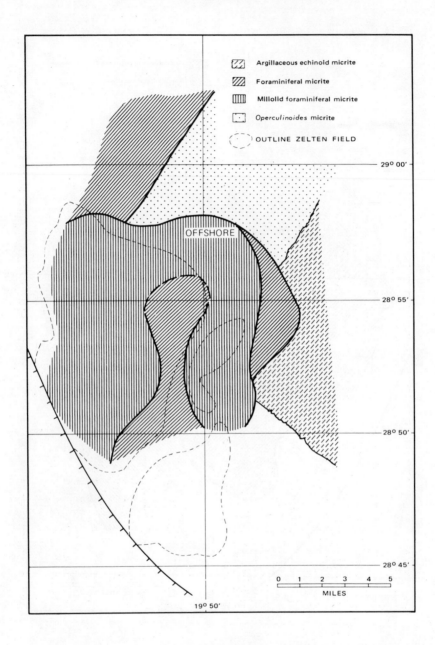

FIG. 21—Facies map of horizon g. Offshore foraminiferal micrite is being deposited over entire Zelten area. Foraminifers representing deeper water conditions, alveolinids and *Operculinoides,* are more common. In overlying Meghil Member *Discocyclina,* represented by small species, and planktonic foraminifers are abundant, climaxing this transgressive sequence.

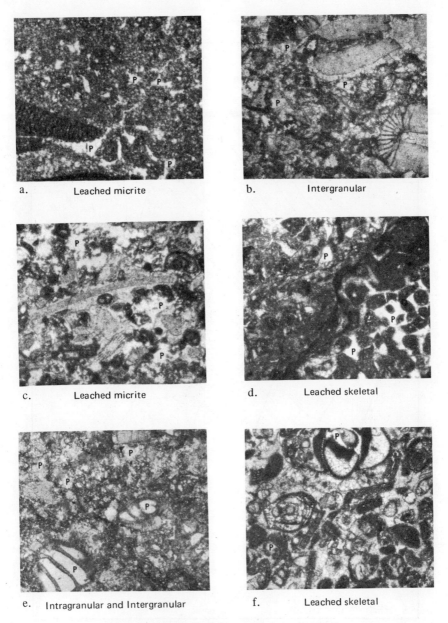

a. Leached micrite	b. Intergranular
c. Leached micrite	d. Leached skeletal
e. Intragranular and Intergranular	f. Leached skeletal

ZELTEN POROSITY TYPES (ALL x 21)

FIG. 22—**a.** C33-6, depth 5,843 ft. Recrystallized and leached micrite, *Discocyclina*-coralgal micrite. Very fine pinpoint vug porosity associated with microspar is typical of this process. **b.** C54-6, depth 5,855 ft. Primary intergranular porosity, *Discocyclina*-foraminiferal calcarenite. Partial plugging of porosity is evidenced by calcite overgrowths on echinoid fragments (upper right) and filling of foraminifer chamber (lower right). **c.** C44-6, depth 5,666 ft. Leached micrite porosity, foraminiferal calcarenite and micrite. Calcite overgrowths on echinoid fragments suggest period of spar filling of primary vugs which were associated with skeletal material prior to leaching of fine micrite particles. **d.** C33-6, depth 5,667 ft. Leached skeletal porosity, coralgal micrite. Corals, such as *Millepora* (lower right) in this figure, are particularly susceptible to leaching. **e.** C33-6, depth 5,792 ft. Primary intragranular and intergranular porosity, foraminiferal calcarenite and micrite. Intragranular porosity is shown within foraminiferal tests; intergranular porosity is between foraminifers and subsequently has been partly plugged by calcite (light colored). **f.** C54-6, depth 5,859 ft. Leached skeletal porosity, foraminiferal calcarenite and micrite. This section indicates early history of spar cementation of calcarenite and later removal of miliolid shell material.

213

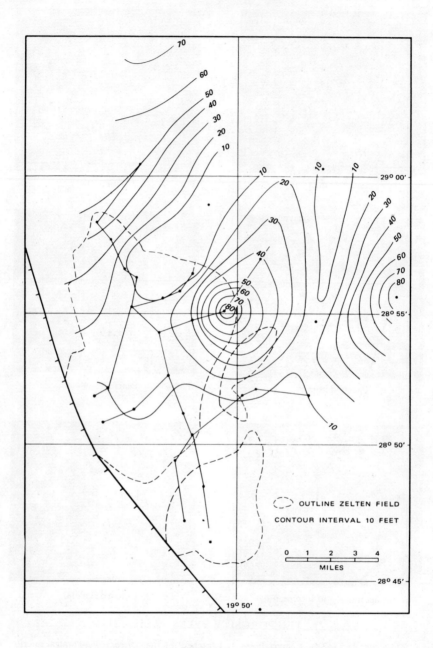

Fig. 23—Isopach of highly porous *Discocyclina*-foraminiferal calcarenite facies.

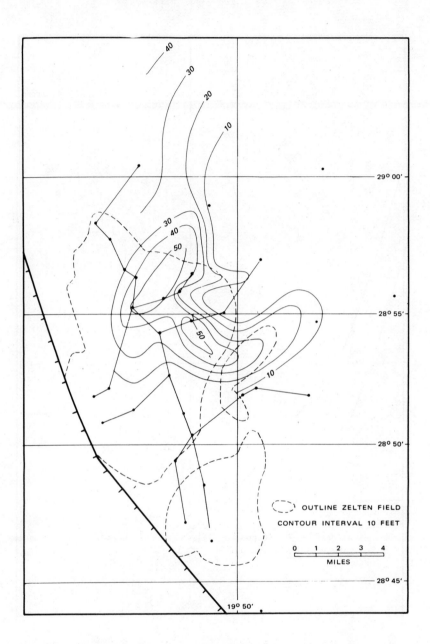

FIG. 24—Isopach of highly porous coralgal micrite facies.

D. G. Bebout and Charles Pendexter

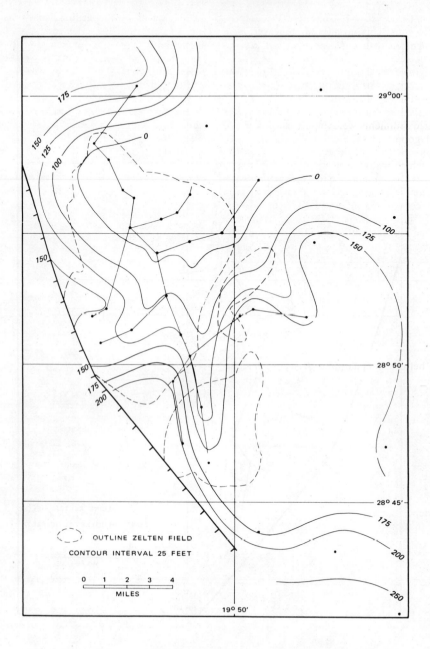

FIG. 25—Isopach of dolomite in lower Zelten Member.

gal micrite, *Discocyclina*-foraminiferal calcarenite, and argillaceous echinoid micrite), however, have been recognized with confidence within the dolomite sequence.

The origin of the dolomite remains in question, but evidence from this study suggests two possibilities. First, the location of the dolomite south of the coralgal and calcarenite trend suggests that the dolomitized sediments were deposited on a broad shelf or lagoon and were dolomitized early after deposition—perhaps at the time of maximum regression, which would correspond approximately to the top of the dolomite. In this case the location of the dolomite is facies or environmentally controlled. Second, cross sections on a structural datum show that the dolomite is restricted to the area south of the Zelten structure and that the dolomite thins and pinches out on that structure. Thus, this alternative interpretation suggests that the Zelten structure was formed first and that later dolomitizing solutions came from the south, flowed along on top of the tight argillaceous Cra Member, and dolomitized all of the carbonate except that high on the structure.

Conclusions

For a better understanding of secondary carbonate porosity it is necessary to become familiar with the controls of porosity development, such as depositional facies, sedimentary structures, chemistry of pore fluids, time of porosity development, and the thermodynamic and kinetic processes involved. This study of the secondary porosity of the Zelten field is only one step toward understanding these relations in the evolution of predictive models.

This study shows that the Zelten porosity is related to depositional facies and, perhaps more specifically, that porosity is inversely proportional to the degree of postdepositional packing. Persistence of depositional fabrics results in continuing permeability for the movement of leaching groundwater solutions. From the position of the porosity in the most regressive stage of a cycle, it is probable that the secondary porosity developed very early after deposition by exposure to freshwater leaching and alteration. Solution by subsurface water at a later stage, however, may have taken place also.

References Cited

Banner, F. T., and G. V. Wood, 1964, Recrystallization in microfossiliferous limestones: Geol. Jour., v. 4, pt. 1, p. 21-34.

Bathurst, R. G. C., 1958, Diagenetic fabrics in some British Dinantian limestones: Liverpool and Manchester Jour. Geology, v. 2, p. 11-36.

———— 1959, Diagenesis in Mississippian calcilutites and pseudobreccias: Jour. Sed. Petrology, v. 29, p. 365-376.

Bowerman, J. N., 1967, Petroleum developments in North Africa in 1966: AAPG Bull., v. 51, p. 1564-1586.

Kemper, E., 1966, Beobachtungen an obereozänen Riffen am Nordrand des Ergene-Beckens (Turkisch-Thrazien): Neues Jahrb. Geologie u. Palaontologie Abh., v. 125, p. 540-553.

Lucia, F. J., 1962, Diagenesis of a crinoidal sediment: Jour. Sed. Petrology, v. 32, p. 848-865.

Murray, R. C., 1960, Origin of porosity in carbonate rocks: Jour. Sed. Petrology, v. 30, p. 59-84.

Nagappa, Y., 1959, Foraminiferal biostratigraphy of the Cretaceous-Eocene succession in the India-Pakistan-Burma region: Micropaleontology, v. 5, p. 145-192.

Nicod, M.-A., 1973, Petroleum developments in North Africa in 1972: AAPG Bull., v. 57, p. 1984-2007.

Terry, C. E., and J. J. Williams, 1969, The Idris "A" bioherm and oilfield, Sirte basin, Libya; its commercial development, regional Palaeocene geologic setting and stratigraphy, *in* The exploration for petroleum in Europe and North Africa: London, Inst. Petroleum, p. 31-48.